ANIMAL SCIENCE, ISSUES AND PROFESSIONS

PRIMATES

CLASSIFICATION, EVOLUTION AND BEHAVIOR

Animal Science, Issues and Professions

Additional books in this series can be found on Nova's website under the Series tab.

Additional e-books in this series can be found on Nova's website under the e-book tab.

ANIMAL SCIENCE, ISSUES AND PROFESSIONS

PRIMATES

CLASSIFICATION, EVOLUTION AND BEHAVIOR

EVERETT F. HUGHES
AND
MADISON E. HILL
EDITORS

Nova Science Publishers, Inc.
New York

For permission to use material from this book please contact us:
Telephone 631-231-7269; Fax 631-231-8175
Web Site: http://www.novapublishers.com

NOTICE TO THE READER

Additional color graphics may be available in the e-book version of this book.

Library of Congress Cataloging-in-Publication Data

ISBN: 978-1-62081-498-7

Published by Nova Science Publishers, Inc. † New York

Contents

Preface

In this book, the authors present topical research in the study of the classification, evolution and behavior of primates. Topics discussed in this compilation include the analysis of the attachments, relative weights, and innervation of the lower limb muscles of various apes and modern humans; Alzheimer pathology in non-human primates and its pathophysiological implications; the evolutionary transformations and adaptations related to the different locomotor modes developed by hylobatids, orangutans, gorillas, chimpanzees and modern humans; neurobiological correlates of behavioral and cognitive performance in nonhuman primates; and assessing the cognitive capacities of various nonhuman primate species.

Chapter I - In this chapter the authors report the results of an analysis of the attachments, relative weights, and innervation of the lower limb muscles of apes and modern humans. Within the hominoid specimens dissected by us the scansorius, articularis genus, adductor minimus, fibularis tertius, quadratus plantae, opponens digiti minimi and opponens hallucis are usually missing in hylobatids, which usually have 60 lower limb muscles. The articularis genus, adductor minimus, fibularis tertius, quadratus plantae and opponens hallucis are usually missing in orangutans, which thus usually have a total of 62 lower limb muscles. The scansorius, gemellus superior, fibularis tertius, plantaris and opponens hallucis are usually absent in gorillas, which usually have 62 lower limb muscles. The scansorius, articularis genus, adductor minimus, fibularis tertius, abductor metatarsi quinti, and opponens hallucis are usually missing in chimpanzees, which usually have 61 muscles. Lastly, in modern humans the scansorius, ischiofemoralis, abductor metatarsi quinti, adductor minimus, opponens digiti minimi and opponens hallucis are usually missing, so there are usually 61 muscles. Thus modern human bipedalism is not associated with an increase in the number of lower limb muscles or with the presence of muscles that are missing in other closely-related primates, but instead to a change in the general configuration, attachments and/or relative mass of the striated muscles of the lower limb.

Chapter II - Alzheimer's disease (AD) is commonly believed exclusive to humans, even though amyloid accumulation has been detected in senile individuals of all the sub-orders, families, genera and species of non-human primates studied to date. Indeed, in some species, all senile individuals show such deposits. However, only rarely are they associated with neuropathological abnormalities. Further, the highly phosphorylated tau protein seen in humans with AD is only very rarely encountered in non-human primates. In fact it is only detected with any regularity in aged mouse lemurs (*Microcebus murinus*), and the changes related to its accumulation appear to be of very little significance. This chapter discusses the pathophysiological meaning of the results obtained in the neuropathological study of the brains of higher (*Chlorocebus aethiops*, *Macaca mulata* and *Macaca arctoides* among others) and lower (Microcebus murinus among others) non-human primates. In macaques, the existence of a possible continuum between the "normal" ageing process, "normal" ageing with no profound neuropathological or cognitive-behavioural changes, and "pathological ageing" (or "Alzheimer-type ageing") is contemplated. The study of non-human primate species is of prime importance for understanding AD but needs to be performed using precise, standardised protocols.

Chapter III - The rotator cuff muscles (subscapularis, supraspinatus, infraspinatus and teres minor) are the principal stabilizers of the glenohumeral joint in primates. This function is particularly important in hominoids due to their orthograde corporal pattern, characterized by a dorsal position of the scapulae, a high mobility of the glenohumeral joint and the presence of tensile forces that affect the shoulder region during locomotion. These functional and physiological conditions imply an special morphology of the rotator cuff muscles in hominoids, which present an increased global mass and important differences in the proportion of the different muscles with respect to the more quadrupedal pronograde primates. Although sharing some similarities, the different general that form the Hominoidea have developed different types of locomotion such as brachiation (*Hylobates*), arm-swinging and vertical climbing (*Pongo*), knuckle-walking (*Gorilla* and *Pan*) and bipedalism (*Homo*). Thus, the morphology of the rotator cuff muscles in these genera is also related to the functional requirements of these types of locomotion. In this chapter the authors will discuss the anatomical and functional aspects of the rotator cuff muscles in hominoids by focusing on the evolutionary transformations and adaptations related to the different locomotor modes developed by hylobatids, orangutans, gorillas, chimpanzees and modern humans.

Chapter IV - Animals in experiments are traditionally grouped by experimental treatment. Although this is a valuable way to differentiate the groups, alternatively, groups can be distinguished based on cognitive performance. Performance based analysis can yield valuable insights, corresponding to behavior and/or cognition, that might not otherwise be observed. As an example of such an analysis, the authors discuss a cohort of elderly female rhesus macaques who participated in a spatial food port maze navigational test. Circadian activity and pharmacological MRI (phMRI) were assessed in these monkeys in vivo and radioligand binding was assessed in post-mortem tissue. Based on cognitive performance in the spatial maze, the cohort of monkeys was divided into Good Spatial Performers (GSP) and Poor Spatial Performers (PSP). GSP animals were more active during the day and less active at night compared to PSP animals. In addition, GSP animals had a higher percentage change in blood-oxygen-level-dependent (BOLD) signal after a scopolamine challenge, a non-specific muscarinic receptor antagonist, compared to PSP animals. Post-mortem radioligand experiments demonstrated that hippocampal muscarinic type 1 (M1) maximum receptor binding and receptor binding affinity, hypothesized to have an integral role in spatial learning and memory, was significantly greater in the hippocampus of GSP than PSP animals. In contrast to the hippocampus, M1 receptor binding was not significantly different in the prefrontal cortex (PFC). Muscarinic type 2 (M2) maximum receptor binding and receptor binding affinity were not significantly different between the two groups in either brain region. Finally, there were positive correlations between circadian measures and the percentage change in BOLD signal following the scopolamine challenge, as well as M1 receptor binding measures. Thus, GSP animals sleep more and have anenhanced M1 receptor function. These data demonstrate the close relationship between BOLD signal changes, circadian activity, and M1 receptor binding parameters. Distinguishing groups based on cognitive or behavioral performance is valuable for studying neurobiological correlates of performance in nonhuman primates.

Chapter V - One of the dominant research areas in behavioral primatology pertains to assessments of the cognitive capacities of various nonhuman primate species. Experiments in this area have shown that primates have many skills that approximate those found in humans. These tests include both manual and computerized assessments, and often the format in which tests are presented can impact the learning and behavioral responding that are produced by subjects. In this chapter, the authors outline two case studies that compare different methodologies and how those methodologies might contribute to differential learning and cognitive performance. In the first, rhesus monkeys responded during a learning task using either joysticks or touch-screens. The task assessed whether the monkeys responded according to stimulus-response associative processes or by using more relational rules about the task. Performance was similar using both input methods and suggested a more associative response pattern. In the second case, chimpanzees and monkeys learned to sequence stimuli using either a computerized task or through selection of three-dimensional stimuli presented in a manual test. Here, the manual test produced faster acquisition and better performance than the computer task, indicating that in some contexts a hands-on, real world format may produce improved cognitive performance.

In: Primates
Editors: E. F. Hughes and M. E. Hill
ISBN: 978-1-62081-498-7

Chapter I

Comparative Anatomy of the Lower Limb Muscles of Hominoids: Attachments, Relative Weights, Innervation and Functional Morphology

Eva M. Ferrero[1,2,3], Juan Francisco Pastor[1], Felix De Paz Fernandez[1], Mercedes Barbosa Cachorro[1], Rui Diogo[2] and Bernard Wood[3]

[1] Department of Anatomy,Valladolid University, Spain
[2] Department of Anatomy, Howard University, Washington, D. C., US
[3] Center for the Advanced Study of Hominid Paleobiology, George Washington University, Washington, D. C., US

Abstract

In this chapter we report the results of an analysis of the attachments, relative weights, and innervation of the lower limb muscles of apes and modern humans. Within the hominoid specimens dissected by us the scansorius, articularis genus, adductor minimus, fibularis tertius, quadratus plantae, opponens digiti minimi and opponens hallucis are usually missing in hylobatids, which usually have 60 lower limb muscles. The articularis genus, adductor minimus, fibularis tertius, quadratus plantae and opponens hallucis are usually missing in orangutans, which thus usually have a total of 62 lower limb muscles. The scansorius, gemellus superior,. fibularis tertius, plantaris and opponens hallucis are usually absent in gorillas, which usually have 62 lower limb muscles. The scansorius, articularis genus, adductor minimus, fibularis tertius, abductor metatarsi quinti, and opponens hallucis are usually missing in chimpanzees, which usually have 61 muscles. Lastly, in modern humans the scansorius, ischiofemoralis, abductor metatarsi quinti, adductor minimus, opponens digiti minimi and opponens hallucis are usually missing, so there are usually 61 muscles. Thus modern human bipedalism is not associated with an increase in the number of lower limb muscles or with the presence of muscles that are missing in other closely-related primates, but instead to a change in the general configuration, attachments and/or relative mass of the striated muscles of the lower limb.

Introduction

One of the first detailed anatomical descriptions of an ape, presumably a chimpanzee, by Nicolaas Tulp, a Dutch physician, was published in 1641 ("Medicae Obserbationes"). Shortly afterwards the English surgeon Tyson published his comparative study of chimpanzees and modern humans (Tyson, 1699). Subsequently our understanding of the anatomy of chimpanzees was expanded by Traill (1821), Vrolik (1841) and Owen (1846). The first dissection of a gibbon is attributed to Le Comte (1697), but the first detailed anatomical description of a lesser ape was that of Kohlbrügge (1890-1891). The earliest study of the anatomy of the orangutan is by Camper (1779, 1782), with additional information being provided by Owen (1830-1831, 1843). The latter author also undertook the first detailed dissection of a gorilla (Owen, 1849, 1859, 1865), and the first authors to recognize the differences between the two genera of African apes were Savage and Wyman (1847). Comparative studies of the apes (Duvernoy, 1855-1856; Huxley, 1864, and Owen 1868) were conducted in parallel with studies that compared the modern human phenotype with that of other primates (e.g., Le Double, 1897; Testut, 1884). Adaptations to the bipedal type of locomotion seen in modern humans include modifications to the muscles of the hind limb. These include the relative enlargement of the superficial gluteal muscles (especially the gluteus maximus), the quadriceps femoris and tensor fascia latae, and the insertion of the flexor digitorum lateralis onto the hallux as the 'flexor hallucis longus' of modern human anatomy. The hard-tissue adaptations include shortening and anteroposterior expansion of the ilium, a more posterior orientation of the acetabulum and a change in the angle of femoral torsion so that both condyles of the modern human knee are parallel and perpendicular to the frontal plane. The physiological valgus angle of the modern human knee is *c*.10° whereas in apes it is either smaller or non-existent. This angulation aligns the leg with a vertical axis, a requirement for efficient standing and walking. The tibio/fibular-talar joint tilt angle differs in modern humans and in the apes; the range of dorsiflexion is in general greater in the apes than in modern humans (Stern and Susman, 1983; Latimer, Ohman, Lovejoy, 1987). The foot of non-human primates has an opposable big toe that is separated from the others digits; this provides greater mobility and the means to grip. White and Suwa (1987) and Gebo (1989) found a number of differences between modern humans and apes, including that in modern humans the bones of the foot are shorter. Current studies such as those of Vereecke et al. (2005ab), Vereecke and Aerts (2008), and Potau (2011) are re-examining the musculature of the apes using a range of techniques that include electromyography, fast cameras, platforms that measure pressure, or studies of myosin heavy chain, which provide new information on the muscular system. However, a detailed comparison of the lower limb musculature of higher primates has not been undertaken. Thus, this chapter reviews, describes, compares and analyzes the muscular system of the lower limb of all major groups of extant hominoids (i.e., hylobatids, orangutans, gorillas, chimpanzees/bonobos and modern humans).

Comparative Analysis of the Origins, Insertions and Function of Hominoid Hindlimb Muscles

In this Section we provide a detailed comparison of the lower limb muscles of extant hominoids. These are based on a review of the literature and information derived from our dissections of three hylobatids (*Hylobates lar* adult male: HU HL1; *H. lar* juvenile female: GWU HL1; *H. (nomascus) gabriellae* adult male: VU HG1), two orangutans (*Pongo pygmaeus* adult male: GWU PP1; *P. pygmaeus* adult female: VU PP1), one gorilla (*Gorilla gorilla* adult female: VU GG1), three chimpanzees (*Pan troglodytes* adult male: VU PT1; *P. troglodytes* adult female: GWU PT1; *P. troglodytes* adult female: GWU PT2) and one modern human (i.e., *Homo sapiens*) (adult female: HU HS1E). The individual hindlimb muscles are reviewed in four regional (pelvis to thigh and leg, thigh, leg and foot) categories. The origins and insertions of each muscle of each genus are summarized in Tables 1-4. The percentage values of the overall hindlimb muscle mass are shown in Tables 5 and 6.

Muscles Connecting the Pelvis to the Thigh and Leg (Table 1)

The *iliopsoas* complex is formed by three muscles, the iliacus, psoas major and psoas minor. Contrary to the condition in the apes, the psoas minor is said to be usually absent in modern humans (Aiello and Dean, 1990). The iliopsoas is the main hip flexor muscle in modern humans, acting as a flexor and stabilizer of the thigh at the hip; it is a postural muscle that helps control the deviation of the trunk during standing (Moore and Dallas, 2002). Both the iliacus and psoas major insert onto the femur, whereas the psoas minor inserts onto both the femur and the pelvic bone. Thus the latter muscle can flex the thigh and the trunk, or it can act on the lumbar region. In apes the function of the iliopsoas is similar to that in modern humans. In all hominoids the iliacus originates from the iliac fossa. Contrary to the condition in the apes, in modern humans the iliopsoas muscle has partially originates from the ala of the sacrum. Both modern humans and *Gorilla* share a partial origin from the lumbodorsal fascia. Apes have an insertion onto the lesser trochanter; its medial surface (*Hylobates* and *Pan*), its lateral surface (*Pongo*), or onto both surfaces (*Gorilla*). In modern humans there is an insertion onto both the lesser trochanter and onto the tendon of the psoas major (Hepburn, 1892; Sigmon, 1974; Kohlbrügge, 1890-91; Ferrero, 2011; our study). The iliacus muscle is innervated by the femoral nerve (L_2-L_4) in modern humans and in all apes (Raven, 1950; Warwick and Williams, 1973; Sigmon, 1974; Gibbs, 1999). All hominoids have an origin of the psoas major from the sides of the bodies and costiform processes of the lumbar vertebrae, as well as from the last thoracic vertebra. Only modern humans have an origin from L5, only *Pan* has an origin from the ribs, particularly the 12th, and only *Gorilla* has an origin from the ilium. In all hominoids the insertion of the psoas major is onto the lesser trochanter of the femur, and only in *Pan* does the insertion extend distally to the lesser trochanter (Hepburn, 1892; Sigmon, 1974; Kohlbrügge, 1890-91; Ferrero, 2011; our study).

Table 1. Presence/absence of specific origins and insertions for the muscles that connect the pelvis to the thigh and the leg in different hominoid taxa

Origin	*Hy*	*Pg*	*Gll*	*Pn*	*Hm*	Muscle	Insertion	*Hy*	*Pg*	*Gll*	*Pn*	*Hm*
Iliac fossa	*	*	*	*	*	*Iliacus*	Lesser trochanter, medial surface	*		*	*	
Sacral ala					*		Lesser trochanter, lateral surface		*	*		
Thoracolumbar fascia			*		*		Lesser trochanter, distal region					*
							Psoas major tendon					*
Lateral lumbar vertebral body	*	*	*	*	*	*Psoas major*	Lesser trochanter	*	*	*	*	*
Transverse processes of lumbar vertebrae	*	*	*	*	*		Lesser trochanter, distal region				*	
T12	*	*	*	*	*							
L5					*							
S1	*		*	*								
Intervertebral disc	*	*		*								
Twelfth rib				*								
Ilion			*									
L1 antero-lateral surface	*	*		*		*Psoas minor*	Iliopubic eminence	*	*	*	*	*
L1 lateral surface					*		Pectineal line	*	*	*	*	*
L2 antero-lateral surface		*		*								
Last thoracic vertebra	*	*		*								

Table 1. (Continued)

Origin	*Hy*	*Pg*	*Gll*	*Pn*	*Hm*	Muscle	Insertion	*Hy*	*Pg*	*Gll*	*Pn*	*H*
T11antero-lateral surface			*			*Psoas minor* (cont.)						
T12 antero-lateral surface			*									
T12 lateral surface					*							
T13 antero-lateral surface			*									
Intervertebral disc	*	*		*	*							
Posterior iliac crest	*	*	*	*		*Gluteus maximus*	Posterior femur surface	*	*		*	
Posterior gluteal line of ilium					*		Lateral femur surface	*	*	*	*	
Posterior superior iliac spine	*		*				Gluteal tuberosity	*	*		*	*
Posterior sacral surface	*	*	*	*	*		Trochanteric fossa				*	
Posterior coccyx surface	*	*	*	*	*		Iliotibial tract	*	*	*	*	*
Ischial tuberosity	*	*	*	*			Vastus lateralis aponeurosis	*				
Sacrotuberous ligament	*	*	*	*	*							
Sacrospinous ligament		*										
Gluteus medius fascia	*	*		*								
Thoracolumbar fascia	*	*	*	*								
Posterior ilium surface	*	*	*	*	*	*Gluteus medius*	Greater trochanter, lateral surface	*	*	*	*	*
Posterior sacral surface			*				Vastus lateralis aponeurosis		*			

Origin	*Hy*	*Pg*	*Gll*	*Pn*	*Hm*	Muscle	Insertion	*Hy*	*Pg*	*Gll*	*Pn*	*Hm*
Gluteal fascia	*	*		*		*Gluteus medius* (cont.)						
Fascia lata			*	*								
Posterior ilium surface	*		*	*		*Gluteus minimus*	Greater trochanter, anterior surface	*	*	*	*	*
Posterior-lateral ilium surface	#	#			*		Greater trochanter, lateral surface		*	*		
Lateral ilium edge				*			Femoral neck			*		
Great sciatic notch	*	*		*			Femoral head capsule		*			
Ischial body		*		*								
Ischial spine		*		*								
Sacrospinous ligament		*										
Fascia lata			*									
Posterior-lateral ilium surface	*	*	*	#		*Scansorius*	Greater trochanter, anterior surface	*	*			
Antero-lateral ilium surface		*					Lateral proximal femur			*	#	
Anterior superior iliac spine		#										
Anterior inferior iliac spine		#										
Fascia lata			*									
Ischial tuberosity	*	*	*	*		*Ischiofemoralis*	Femoral diaphysis, lateral surface	*	*	*	*	
							Vastus lateralis aponeurosis	*	*	*	*	

Table 1. (Continued)

Origin	*Hy*	*Pg*	*Gll*	*Pn*	*Hm*	Muscle	Insertion	*Hy*	*Pg*	*Gll*	*Pn*	*Hm*
Anterior superior iliac spine	*	*	*	*	*	*Tensor fasciae latae*	Iliotibial tract	*	*	*	*	*
Iliac crest, ventral region					*		Fascia of vastus lateralis			*		
Gluteal fascia	*			*								
Ischial spine				*	*	*Gemellus superior*	Trochanteric fossa	*	*		*	*
Ischial body	*	*										
Sacrospinous ligament				*								
Ischial tuberosity	*	*		*	*	*Gemellus inferior*	Trochanteric fossa	*	*	*	*	*
Ischial body			*	*								
Obturator foramen, outer surface	*	*	*	*	*	*Obturator externus*	Trochanteric fossa	*	*	*	*	*
Obturator membrane	*	*	*	*	*		Lesser trochanter				#	
							Femoral neck, proximal surface			*	*	
Obturator foramen, inner surface	*	*	*	*	*	*Obturator internus*	Trochanteric fossa	*	*	*	*	*
Obturator membrane	*	*	*	*	*		Medial surface of greater trochanter					*
Anterior sacral surface	*	*	*	*	*	*Piriformis*	Greater trochanter	*	*	*	*	*
Lateral sacral surface	*	*	*									
Greater sciatic notch, proximal surface	*	*		*								
Lateral coccyx surface			*									

Origin	*Hy*	*Pg*	*Gll*	*Pn*	*Hm*	Muscle	Insertion	*Hy*	*Pg*	*Gll*	*Pn*	*Hm*
Sacrotuberous ligament					*	*Piriformis* (cont.)						
Ischial tuberosity, lateral surface	*	*		*	*	*Quadratus femoris*	Intertrochanteric crest	*	*	*	*	*
Ischial tuberosity, anterior surface		*		*			Lesser trochanter	*	*	*	*	
Ischial body			*				Greater trochanter			*	*	
							Greater/lesser trochanters, posterior surface		#		#	
							Third trochanter		*			
							Gluteal tuberosity		*			
							Ischiofemoral ligament		*			

In this and the other tables, *Hy*, *Pg*, *Gll*, *Pn* and *Hm* mean *Hylobates*, *Pongo*, *Gorilla*, *Pan* and *Homo*; * indicates the presence of muscle origin or insertion that is observed in our specimens and in the literature; # indicates the presence of muscle origin or insertion observed by other authors but not by us; + indicates the presence of muscle origin or insertion observed by us and not previously reported in the literature.

The psoas major muscle is innervated by branches of the lumbar plexus (L2-L3) in modern humans, while in Asian and African apes it is innervated by the first two or three lumbar nerves and the femoral nerve (Champneys, 1872; Hepburn, 1892; Sonntag, 1923; Raven, 1950; Sigmon, 1974, Gibbs, 1999). The psoas minor is usually absent in modern humans (Aiello and Dean, 1990) but it was found in all ape specimens dissected by us. Its origin is highly variable, the most common origin being the anterolateral aspect of L1 and the last thoracic vertebra (*Hylobates*, *Pongo* and *Pan*). In *Gorilla* it originates from the anterolateral aspects of the bodies of T11, T12 and T13. Modern humans also have an origin from the lateral aspect of T12, and, as in apes except for *Gorilla*, also from the lumbar vertebrae, particularly the lateral aspect of L1; the psoas minor of apes thus has a more lateral origin than is the case in modern humans. Among the apes, *Gorilla* has a more cranial origin, reaching the thoracic vertebrae. In all hominoids the insertion of psoas minor is onto the iliopubic eminence and the dentate line (Sigmon, 1974; Kohlbrügge, 1890-91; Ferrero 2011; our study), although Hepburn (1892) stated that the insertion is more anterior in *Pan* and modern humans than in *Pongo*, an observation that was not corroborated in our dissections. Innervation of the psoas minor muscle is by the first lumbar nerve (Champneys, 1872; Hepburn, 1892), and sometimes also by the twelfth thoracic nerve in Asian apes and *Pan* (Hepburn, 1892; Sigmon, 1974). In *Gorilla* it is by branches of the femoral nerve (Raven, 1950) as well as by L1 and occasionally also by T12 (Gibbs, 1999).

The *gluteal muscles* are arranged in two planes: superficial (gluteus maximus, gluteus medius, gluteus minimus, scansorius, ischiofemoralis and tensor fascia latae) and deep (gemellus superior, gemellus inferior, obturator externus, obturator internus, piriformis, and quadratus femoris). The function of the deep muscles is mainly to laterally rotate the thigh. Some authors (e.g., Aiello and Dean, 1990) consider this function to be important for bipedal locomotion because in modern humans these muscles are functionally independent and relatively short. The *gluteus maximus* is a flat muscle which is thinner proximally than distally (Hepburn, 1892; Owen, 1830-1831; Sigmon, 1974; Sonntag, 1924; Zihlman and Brunk, 1979; Ferrero, 2011; this study). All hominoids have an origin from the posterior surface of the sacrum, coccyx and sacrotuberous ligament. Contrary to the condition in modern humans, in apes there is an origin from the thoracolumbar fascia, the ischial tuberosity and the posterior iliac crest. Contrary to apes, in modern humans the gluteus maximus originates from the posterior gluteal line of the ilium. In *Pongo* the muscle is partially originates from the sacrospinous ligament.

According to some authors (Hamada, 1985; Sigmon, 1974; Stern, 1972) the insertion of gluteus maximus is usually more distal in apes than in modern humans; this is due to the fact that these authors consider the ischiofemoralis as a distal portion of the gluteus maximus. In all hominoids the gluteus maximus has an insertion onto the iliotibial tract, and in all apes it has an insertion onto the lateral aspect of the femur. *Gorilla* does not have an insertion in the region of the gluteal tuberosity, *Pan* has an insertion onto the femoral trochanteric fossa, and *Hylobates* has an insertion onto the aponeurosis of the vastus lateralis. Our observations agree in general with previous reports (Beddard, 1893; van den Broek, 1914; Brown, 1983; Hamada, 1985; Hepburn, 1892; Kohlbrügge, 1890-91; MacDowell, 1910; Miller 1945; Mysberg, 1917; Ranke, 1897; Robinson et al, 1972; Sigmon, 1974; Appleton and Ghey, 1929; Boyer, 1935; Chapman, 1880). However, they contrast with former descriptions of gluteus maximus in *Pan*. According to Champneys (1872) in *Pan* the gluteus maximus shares an origin with the long head of the biceps femoris, while in our specimens instead it shares an origin with the ischiofemoralis. The gluteus maximus is innervated by the inferior gluteal nerve (L5-S2) in modern humans, *Gorilla* and *Pongo* (Raven, 1950; Sigmon, 1974; Gibbs 1999). In modern humans the muscle is an important extensor of the thigh and is the largest hip muscle; this is not the case in apes, according to Marzke et al. (1988). In modern humans the muscle also acts as a lateral rotator of the thigh; during modern human bipedal gait it helps to balance and control the trunk during activities involving the upper extremity (Marzke et al., 1988), being used when extra strength is needed (e.g., when running, or rising from a sitting position). Some authors have reported that in modern humans the muscle has a larger area of origin than the other gluteal muscles (Aiello and Dean, 1990), but this is not corroborated by our dissections nor by other authors (Hepburn, 1892, Sonntag, 1924). In our specimens the gluteus medius has the largest area of origin.

The *gluteus medius* originates from the posterior surface of the ilium in all hominoids, as well as from the posterior aspect of the sacrum in gorillas, and from the fascia lata in African apes. In all hominoids it inserts onto the lateral aspect of the greater trochanter, with *Pongo* having an extra insertion onto the aponeurosis of the vastus lateralis. Previous authors described a fusion of the origins of the gluteus medius and piriformis in hylobatids (Beddard, 1893) and orangutans (Boyer, 1935), but our dissections do not corroborate this observation. Some authors also described a similar fusion in *Pan* (Champneys, 1872; Dwight, 1895; Hepburn, 1892), but we did not see this in our dissected chimpanzees. Innervation of the gluteus medius is by the superior gluteal nerve (L_4-S_1) in modern humans as in the apes (Raven, 1950; Sigmon, 1974; Gibbs, 1999). In modern humans, the gluteus medius acts as a major abductor of the thigh, along with the gluteus minimus, being also a medial rotator of the thigh. According to Aiello and Dean (1990) the situation of this muscle on the lateral aspect of the hip joint and its role as a pelvis stabilizer are uniquely modern human features related to bipedal locomotion.

In modern humans the *gluteus minimus* originates from the postero-lateral surface of the ilium, in *Pan* from the ilium, in *Gorilla* from the fascia lata, and in *Pongo* from the sacrospinous ligament. In Asian apes there is an origin from the postero-lateral surface of the ilium as in modern humans. In all hominoids the gluteus minimus inserts onto the anterior aspect of the greater trochanter, in *Pongo* it has an additional insertion onto the capsule of the hip joint in the region of the femoral head, and in *Gorilla* it has an addition insertion onto the femoral neck (i.e., in these two genera the insertion is more proximal than in other hominoids) (Champneys, 1872; Raven, 1950; Sigmon, 1974; Beddard, 1893; Hepburn, 1892; Ferrero, 2011; this study). Sigmon (1974) described an origin of the gluteus minimus in hylobatids from the postero-lateral surface of the ilium, extending just distal to the anterior superior iliac spine, near the acetabulum, but in the hylobatids dissected by us the origin is from the posterior surface of the ilium. Sigmon (1974) stated that in orangutans there are two heads of origin for the gluteus minimus, but this statement is not corroborated by our dissections nor by the reports of other authors (Boyer, 1935; Chapman, 1880; Raven, 1950; Sigmon, 1974; Sonntag, 1924) which suggest there is a single origin from the postero-lateral aspect of the ilium. The innervation of the muscle is by the superior gluteal nerve (L_4-S_1) in modern humans as in apes (Raven, 1950; Sigmon, 1974; Gibbs, 1999). The gluteus minimus is mainly an abductor, but it is also a lateral and medial rotator, of the thigh. Due to its location on the lateral aspect of the hip joint, it also plays a role in stabilizing the hip during bipedalism (Aiello and Dean, 1990).

The *scansorius* is usually absent in modern humans, it is occasionally present in African apes and hylobatids (Aiello and Dean, 1990), and it is said to be always present in *Pongo* (Barnard, 1875; Boyer, 1935; Primorse, 1899; Sigmon, 1969; Waterman, 1929). Located laterally to the ilium, it is independent of the gluteus medius and minimus. When present in apes, the scansorius muscle originates from the postero-lateral surface of the ilium, and also its antero-lateral surface in orangutans and the fascia lata in *Gorilla*. In *Pongo*, previous authors (Beddard, 1893; Boyer, 1935; Pearson and Davin, 1921) reported an origin from the anterior superior iliac spine and from the anterior inferior iliac spine, but these attachments were not seen in the orangutans dissected by us. In Asian apes the muscle inserts onto the anterior aspect of the greater trochanter, and in African apes (Beddard, 1893; Blake, 1976; Champneys, 1872; Hepburn, 1892; Wilder, 1862) it inserts onto the lateral aspect of the proximal femur. The scansorius is said to be occasionally found in *Pan* (Beddard, 1893; Blake, 1976; Champneys, 1872; Hepburn, 1892; Wilder, 1862) but it is was not present n the chimpanzees dissected by us. We did, however, find it in *Gorilla,* hylobatids and *Pongo*. Some authors (Beddard, 1893; Boyer, 1935; Pearson and Davin, 1921) have stated that in orangutans the origin of the muscle extends from the anterior superior to the anterior inferior iliac spines, but in the orangutans dissected by us the origin lay between the two spines without reaching either. Also, the scansorius is described as fused proximally with the iliacus in orangutans (Sigmon, 1974), but we did not observe such a fusion. The scansorius is occasionally present in hylobatids (Beddard, 1893; Blake, 1976; Hepburn, 1892). The scansorius muscle is innervated by the superior gluteal nerve in African apes and *Pongo* (Boyer, 1935; Raven, 1950; Sigmon, 1969; Sigmon, 1974). When present, the scansorius is primarily responsible for extending rather than flexing the thigh, while also abducting and medially rotating it (Sigmon, 1974).

Although initially described by some authors as comprising the distal third of the gluteus maximus (Hamada, 1985; Sigmon, 1974; Stern, 1972), the *ischiofemoralis* is now considered to be an independent muscle, particularly in light of the innervation studies of Kaseda et al. (2008). This muscle is usually missing in modern humans but it is particularly developed in *Pongo*. In apes the muscle runs from the ischial tuberosity to the lateral aspect of the femoral shaft and the aponeurosis of the vastus lateralis. In the orangutans dissected by us the ischiofemoralis is distinct from the surrounding muscles, while in *Hylobates*, *Gorilla* and *Pan* it blends with the gluteus maximus. The ischiofemoralis is innervated by the common peroneal component of the sciatic nerve in *Pongo* and *Pan* (Kaseda et al., 2008). The ischiofemoralis is an extensor of the thigh, and for some authors its particular development in orangutans is a reflection of their adaptation to an arboreal life style (Kaseda et al., 2008). However, it should be noted that in the specimens dissected by us the percentage values of the muscle vs. the total weight of the hindlimb musculature are 5.4% for *Pan*, 5.0% for *Pongo*, and 4.2% for *Hylobates*, thus raising doubts about this functional interpretation.

The *tensor fasciae latae* is present in all of the specimens dissected by us, and is reported to be present in all apes except *Pongo*, where it is presumably functionally replaced by the scansorius (Boyer, 1935; Church, 1861-1862; Hepburn, 1892; Owen, 1830-1831; Sigmon, 1974; Sonntag, 1924). In all hominoids dissected by us the tensor fasciae latae originates from the anterior superior iliac spine. In all apes its insertion is onto the distal iliotibial tract. The muscle is innervated by the superior gluteal nerve (L_4-S_1) in modern humans, *Pongo* and *Gorilla* (Raven, 1950; Sigmon, 1974; Gibbs, 1999). In *Pan troglodytes* innervation is by the superior gluteal nerve (Sigmon, 1974), while the inferior gluteal nerve is cited as its nerve supply in *Pan paniscus* (Miller, 1952). In *Hylobates* the mediodorsal part of tensor fasciae latae is innervated by the superior gluteal nerve, and the anterolateral part by the femoral nerve (Sigmon, 1974). In modern humans, the tensor fasciae latae medially rotates and abducts the thigh, and together with the iliotibial tract it maintains the knee in extension. According to Aiello and Dean (1990) in African apes the tensor fasciae latae originates from the fascia covering the gluteal muscles and from the iliac crest and because of that while it acts as a flexor and abductor of the thigh as in modern humans, it flexes rather than extends the thigh. As the muscle also originates from the gluteal fascia in hylobatids, it is possible that it flexes rather than extends the thigh in these apes.

In Asian apes the *gemellus superior* originates from the ischial body, from the ischial spine in both *Pan* and modern humans and, exclusively in *Pan*, it originates from the sacrospinous ligament. The insertion in all taxa is into the trochanteric fossa. The muscle is reported to be absent in *Gorilla* (Gibbs, 1999) and it was not found in the gorilla specimens dissected by us. According to Sigmon (1974) this muscle is usually absent in hylobatids, but in one specimen dissected by Hepburn (1892) and in the hylobatids dissected by us it is present as a small fascicle that has a common insertion with the gemellus inferior, obturator externus and obturator internus into the trochanteric fossa. Sigmon (1974) also stated that the muscle is absent in some *Pongo*, but in the specimens dissected by us it is present and has a common insertion with the obturator internus and the gemellus inferior into the trochanteric fossa. The gemellus superior has been reported to be be usually present in *Pan* (Beddard, 1893; Champneys, 1872; Dwight, 1895; Hepburn, 1892; Sigmon, 1974) and it was present in the chimpanzees dissected by us; its insertion was fused with that of the gluteus minimus, but not with that of the obturator internus. The gemellus superior is innervated by nerves to obturator internus and quadratus femoris in modern humans, in *Pongo* and *Pan* by sacral nerves via the sacral plexus (Champneys, 1871; Sigmon, 1974) and in *Gorilla* by branches of the sacral plexus from the base of the sciatic nerve (Raven, 1950; Gibbs, 1999).

The origin of the *gemellus inferior* in all hominoids except *Gorilla* is from the ischial tuberosity; in *Gorilla* it is from the ischial body. In all hominoids the gemellus inferior inserts onto the trochanteric fossa of the femur, although unlike in modern humans in apes the insertion blends with that of the obturator internus (Boyer, 1935; Hepburn, 1892; Sigmon, 1974; Sonntag, 1924; Beddard, 1893; Champneys, 1872; Gibbs, 1999; Ferrero, 2011; this study). Contrary to what is often stated in the literature, in the *Pongo* and *Pan* specimens dissected by us there is a close association between the gemellus inferior and the quadratus femoris.

In modern humans, the gemellus inferior is innervated by the nerves to obturator internus and quadratus femoris, by sacral nerves in Asian apes and *Pan* (Champneys, 1871; Sigmon, 1974), and occasionally by a branch from a single nerve in *Pan* (Champneys, 1871).

In modern humans the *obturator externus* originates from the external aspect of the obturator membrane and the adjacent bones, runs inferior to the hip joint, wraps posteriorly around the femoral neck and inserts onto the trochanteric fossa.

This circuitous route is the direct result of the habitually extended posture of the hip-joint of upright modern humans. When the thigh is flexed in modern humans, the course of the obturator externus is more direct, and this is also the case in apes in which the hip joint is habitually more flexed. The origin of the muscle in all hominoids is as it is in modern humans (i.e., from the obturator membrane and the margins of the obturator foramen), and it inserts into the trochanteric fossa of the femur. Beddard (1893) describes an additional insertion in *Pan* onto the lesser trochanter, but we did not observe this in the specimens dissected by us.

African apes have an insertion onto the femoral neck adjacent to the trochanteric fossa; this is not seen in Asian apes or modern humans. Some authors have reported that the insertions of the obturator externus and of the obturator internus are often fused in *Pongo* (Beddard, 1893; Hepburn, 1892). This was the case in one of the two orangutans dissected by us; in the other the obturator externus was fused with the gemellus superior and the gemellus inferior. In modern humans the obturator externus muscle is innervated by the obturator nerve (L_3-L_4) and by a named nerve from the sacral plexus (L_5-S_2), but in apes it is innervated exclusively by the obturator nerve (Sigmon, 1974; Raven, 1950; Gibbs, 1999).

The *obturator internus* was found in all of the specimens dissected by us in the form described by previous authors (Beddard, 1893; Sigmon, 1974; Hepburn, 1892; Boyer, 1935; Sonntag, 1924; Gibbs, 1999; Champneys, 1872; Swindler and Wood, 1973).

The only exception is that in the chimpanzees dissected by us the insertion is fused with that of the gemellus inferior only, and not with both this muscle and the gemellus superior, as described by Beddard (1893), Champneys (1872) and Sigmon (1974). Innervation of the obturator internus is by the sacral nerves in modern humans (Sigmon, 1974) and by S1, S2, S3 and possibly L3 in the apes (Raven, 1950; Gibbs, 1999).

In all hominoids the *piriformis* originates from the anterior surface of the sacrum, as well as from its lateral aspect.

Exclusively in *Gorilla* this muscle takes origin from the lateral coccyx surface and from the sacrotuberous ligament in modern humans. In all hominoids the piriformis inserts onto the greater trochanter of the femur (Beddard, 1893; Hepburn, 1892; Sigmon, 1974; Boyer, 1935; Chapman, 1880; Sonntag, 1924; Gibbs, 1999; Ferrero, 2011; this study).

One chimpanzee dissected by us had an origin from the greater sciatic notch, as was described by Chapman (1880) and Sigmon (1974). Innervation of the piriformis muscle is by the obturator nerve (L_3-L_4) and in modern humans by a named nerve from the sacral plexus (L_5-S_2). In the Asian apes and *Pan* it is also supplied by branches of the sacral plexus (Sigmon, 1974), and in gorillas by a nerve arising from S1 and S2.

In hominoids the *quadratus femoris* usually originates from the lateral edge of the ischial tuberosity, but in gorillas it originates instead from the ischial body. Insertion is always onto the intertrochanteric crest of the femur and, with the exception of modern humans, it also inserts onto the adjacent lesser trochanter. Some authors reported that in *Pongo* and *Pan* there is an insertion onto the posterior aspect of the greater and lesser trochanters (Boyer, 1935; Champneys, 1872; Dwight, 1895; Hepburn, 1892), but we did not found such an insertion in our dissections. In the orangutans dissected by us there was instead a partial insertion onto the third trochanter, the gluteal tuberosity and the ischiofemoralis. Such attachments may facilitate the lateral rotation of the thigh, a movement frequently used by orangutans. Boyer (1935) described an origin of the muscle from the medial aspect of the ischial tuberosity in orangutans, and Sonntag (1924) stated that in these apes the muscle merges with the pectineus; neither condition was found in our dissections, nor did we find two insertion sites as described by some authors (Hepburn, 1892; Sonntag, 1924). Innervation of the quadratus femoris is by nerves from the sacral plexus (L_4-S_1) in modern humans, by sacral nerves in the Asian apes and *Pan* (Sigmon, 1974) and by branches of the sacral plexus along with the nerve to the inferior gemellus in gorillas (Raven, 1950; Gibbs, 1999). The quadratus femoris is a lateral rotator of the thigh.

Table 2. Presence/absence of specific origins and insertions for the thigh muscles in different hominoid taxa (see caption of Table 1)

Origin	*Hy*	*Pg*	*Gll*	*Pn*	*Hm*	Muscle	Insertion	*Hy*	*Pg*	*Gll*	*Pn*	*Hm*
Anterior inferior iliac spine		*	*	#	*	*Rectus femoris*	Patella	*	*	*	*	*
Iliac crest, distal region	*			*			Knee joint capsule	*	*	*	*	*
Acetabular fossa, proximal region	*	#	*	*	*							
Femoral diaphysis, proximal ventral surface	*	*	*	*	*	*Vastus intermedius*	Patella	*	*	*	*	*
Femoral diaphysis, proximal medial surface	*	*	*	*			Knee joint capsule	*	*	*	*	*
Femoral diaphysis, proximal lateral surface	*	*		*	*							
Femoral neck			*									
Iliofemoral ligament				#								
Greater trochanter, lateral surface	*	*	*	*	*	*Vastus lateralis*	Patella	*	*	*	*	*
Greater trochanter, distal surface	*	*	*		*		Knee joint capsule	*	*	*	*	*
Linea aspera, lateral ridge	*	*	*	*	*							
Iliofemoral ligament		*										
Linea aspera, medial ridge	*	*	*	*	*	*Vastus medialis*	Patella	*	*	*	*	*
Intertrochanteric line					*		Knee joint capsule	*	*	*	*	*
Femoral neck				+								

Table 2. (Continued)

Origin	*Hy*	*Pg*	*Gll*	*Pn*	*Hm*	Muscle	Insertion	*Hy*	*Pg*	*Gll*	*Pn*	*Hm*
Iliofemoral ligament	*	*		#		*Vastus medialis* (cont.)						
Distal femur			*		*	*Articularis genus*	Knee joint capsule			*		*
Lateral ilium	*	*	*			*Sartorius*	Tibial diaphysis proximal medial	*	*	*	*	*
Anterior ilium	#	*		*			Tibial collateral ligament, anterior		*			+
Anterior superior iliac spine, lateral region	*											
Anterior superior iliac spine, inferior region		+		*								
Anterior superior iliac spine		*			*							
Lesser sciatic notch, superior region					*							
Pubic body	#	*	*	*	*	*Adductor brevis*	Pectineal line		*		*	*
Superior pubic ramus	*	*	*	*			Lesser trochanter, distal region	*	*		*	
Inferior pubic ramus		*	*	*	*		1/3 proximal medial ridge			*	*	*
Pubic body					*	*Adductor longus*	1/2 proximal medial ridge	*	*	*	*	

Origin	*Hy*	*Pg*	*Gll*	*Pn*	*Hm*		Insertion	*Hy*	*Pg*	*Gll*	*Pn*	*Hm*
Superior pubic ramus	*	*	*	*			1/3 proximal medial ridge					*
Pubic tubercle	*	*	+	*			Medial femoral condyle				#	
Inferior pubic ramus	*	*	*	*	*	*Adductor magnus*	1/3 medial femur	*				
Ischial ramus	*	*	*	*	*		Linea aspera, medial ridge	*	*	*	*	*
Ischial tuberosity			*	*	*		Adductor tubercle		*		*	*
							Intercondylar line		*			
							Medial femoral epicondyle		*	*	*	
							Medial femoral supracondylar line					*
							Gluteal tuberosity					*
Superior pubic ramus			*			*Adductor minimus*	Lesser trochanter, lateral surface			*		
Inferior pubic ramus			*		*		Gluteal tuberosity					*
Ischial ramus			+				Linea aspera, proximal region			*		*
Superior pubic ramus	*	*	*	*	*	*Pectineus*	Lesser trochanter, distal region	*	*	*	*	*
Pubic tubercle		*					Proximal femur, dorsal surface	*	*	*	*	
							Pectineal line					*

Table 2. (Continued)

Origin	*Hy*	*Pg*	*Gll*	*Pn*	*Hm*	Muscle	Insertion	*Hy*	*Pg*	*Gll*	*Pn*	*Hm*
Pubic body	#	*	*	*	*	*Gracilis*	Tibial diaphysis, proximal medial surface	*	*	*	*	*
Superior pubic ramus		*	*	*			Tibial diaphysis proximal anterior surface		*		*	
Inferior pubic ramus	*	*	*	*	*		Tibial collateral ligament					+
Ischial ramus		*		*								
Ischial tuberosity	*	*	*	*	*	*Biceps femoris*	Lateral tibial condyle	*	*	*	*	
Linea aspera, lateral ridge	*	*	*	*	*		Tibial tuberosity	*	*	*	*	
							Fibular head	*	*	*	*	*
							Knee joint capsule	*	*			
							Iliotibial tract		*	*	*	
							Fibular collateral ligament		*			
							Lateral femoral epicondyle		*			

Origin	*Hy*	*Pg*	*Gll*	*Pn*	*Hm*	Muscle	Insertion	*Hy*	*Pg*	*Gll*	*Pn*	*Hm*
Ischial tuberosity	*	*	*	*	*	*Semimembranosus*	Tibial medial condyle	*	*	*	*	*
							Tibial collateral ligament	*	*	*	*	*
							Oblique popliteal ligament			*		*
Ischial tuberosity	*	*	*	*	*	*Semitendinosus*	Tibial diaphysis, proximal medial surface	*	*	*		*
							Tibial diaphysis, proximal ventral surface				*	
							Tibial tuberosity, medial region	*	*	+	*	
							Tibial collateral ligament		*			

Muscles of the Thigh (Table 2)

The muscles of the thigh are divided into three groups: an anterior or extensor group (quadratus femoris, sartorius and articularis genus), a medial group (four adductors, pectineus and gracilis) and a posterior or flexor group (biceps femoris, semitendinosus and semimembranosus).

The *quadriceps femoris* actually consists of *five* muscles: rectus femoris, vastus intermedius, vastus lateralis, vastus medialis and articularis genus. All inserting onto the joint capsule of the knee and onto the patella. Only the rectus femoris flexes the thigh at the hip joint; all five muscles extend the leg at the knee joint.

The *rectus femoris* has two heads of origin in modern humans and *Gorilla*: the so-called 'direct' head from the anterior inferior iliac spine and the 'reflected' head from the superior margin of the acetabular fossa. In *Hylobates* there is only a 'reflected' head. In *Pongo* the rectus femoris is described as having two heads in one fifth of the specimens (Boyer, 1935; Hepburn, 1892), but the orangutans dissected by us only had a 'direct' head, but a few fibers of the muscle also originate from the acetabular fossa. In *Pan* the muscle is said to have two heads in one third of the specimens (Dwight, 1895; Hepburn, 1892; Vrolik, 1841), but we only found a 'reflected' head in our chimpanzee dissections. In one of the *Pan* specimens dissected by Beddard (1893) the rectus femoris was blended with the vastus lateralis; such a blending was not found in the chimpanzees dissected by us. Innervation of the rectus femoris is by the femoral nerve (L_2-L_4) in modern humans and *Gorilla* (Raven, 1950; Gibbs, 1999).

According to Aiello and Dean (1990) the origins and insertions of the *vastus intermedius, vastus lateralis* and *vastus medialis* are similar in modern humans and in the apes. The vastus intermedius originates in all hominoids from the ventral aspect of the proximal femoral shaft, extending medially in all apes and laterally in all hominoids except *Gorilla*, the only taxon in our study that has an exclusive origin from the femoral neck. The vastus lateralis originates in all hominoids from the two-thirds of the lateral ridge of the linea aspera and the greater trochanter; in all taxa, except in *Pan*, the attachment extends to the distal region of the greater trochanter. In all hominoids the vastus medialis originates from the medial ridge of the linea aspera, uniquely in modern humans it originates from the intertrochanteric line. Sigmon (1974) stated that chimpanzees have an additional origin from the iliofemoral ligament; we did not see this in our specimens, but we did find an additional origin from the femoral neck that was not described by previous authors. All three muscles are innervated by the femoral nerve (L_2-L_4) in modern humans and in *Gorilla* (Raven, 1950; Gibbs, 1999).

According to Hepburn (1892) the *articularis genu* is present in all apes, but we were only able to confirm its presence in *Gorilla*, in which the configuration was similar to that found in modern humans. In modern humans the articularis genu is responsible for pulling the suprapatellar bursa superiorly during extension of the knee. Innervation of the muscle is by the femoral nerve (L_2-L_4) in modern humans and in the *Gorilla* individual dissected by Raven (1950).

Only in modern humans and *Pongo* does the *sartorius* originate from the anterior superior iliac spine; in the other apes the origin is more caudal. Beddard (1893) and Hepburn (1892) describe an origin from the anterior edge of the ilium, but in the three gibbons dissected by us the origin is from the lateral edge of the ilium, as reported by Sigmon (1974). In one of our orangutans the origin is from the region of the anterior superior iliac spine as described in the literature (Boyer, 1935; Humphry, 1866-1867; Primorse, 1899; Sigmon, 1974; Sonntag, 1924), but in the other orangutan the sartorius originates from the inferior region to the anterior superior iliac spine. The origin in *Gorilla* and *Pan* is similar to that described by previous authors (Beddard, 1893; Chapman, 1879; Hepburn, 1892; Sigmon, 1974; Gibbs, 1999) (i.e., from the lateral edge of the ilium in the former, and from the anterior edge of the ilium and the distal part of the anterior superior iliac spine, in the latter). The insertion in all hominoids is onto the medial aspect of the proximal tibial diaphysis; in *Pongo* it extends onto the anterior surface of the tibial collateral ligament. In *Hylobates* the insertion reaches the distal part of the tibial collateral ligament. In all of the hominoids sartorius is innervated by the femoral nerve (L_2-L_3) (Raven, 1950; Gibbs, 1999; Sigmon, 1974). The sartorius mainly flexes, abducts and laterally rotates the thigh and flexes and medially rotates the leg.

The *adductors of the thigh* fan out from the ischium and pubis and insert onto the posterior aspect of the femur. They all adduct the thigh, but the adductor brevis and adductor longus also flex it and the adductor magnus extends it. In all hominoids except the hylobatids the *adductor brevis* originates from the inferior pubic ramus. In the African apes and modern humans the insertion is onto the proximal third of the medial ridge of the linea aspera of the femur. Some authors describe an origin of the muscle in hylobatids from the body of the pubis (Beddard, 1893; Hepburn, 1892; Sigmon, 1974), but in the gibbons dissected by us the origin was from the superior pubic ramus instead. With respect to *Gorilla* and *Pan*, there are no differences between our observations and those of other authors (Appleton and Ghey, 1929; Beddard, 1893; Champneys, 1872; Dwight, 1895; Hepburn, 1892; Sigmon, 1974, Ferrero, 2011). Innervation of the adductor brevis muscle in modern humans is from the obturator nerve (L_3-L_4) and from the ventral division of the obturator nerve in the apes (Champneys, 1871; Sigmon, 1974; Raven, 1950; Hamada, 1985; Gibbs, 1999).

Only in modern humans does the *adductor longus* originate from the body of the pubis; in the apes it originates from the region of the pubic tubercle. Appleton (1929) stated that in *Hylobates* the insertion is more proximal than in other apes, but in all of the apes dissected by us the insertion is onto the middle third of the medial ridge of the linea aspera of the femur (or its equivalent), and in modern humans it is onto the proximal third of this ridge. Boyer (1935) reported a specimen of *Pongo* in which the muscle arises from the margin of the origin of the pectineus, but this was not the case in the orangutans we dissected. Raven (1950) described the muscle as lying near the pectineus in *Gorilla*; in our gorilla there was an additional origin from the region proximal to the pubic tubercle. Humphry (1866-1867) reported an expansion of the insertion of adductor longus to the medial condyle of the femur in chimpanzees; we did not see this in our apes. Innervation of the muscle is by the obturator nerve (L_3-L_4) in modern humans and from the ventral division of this nerve in apes (Champneys, 1871; Sigmon, 1974; Raven, 1950; Hamada, 1985; Gibbs, 1999).

In all hominoids the *adductor magnus* originates from the inferior pubic ramus and the ischial ramus; only in African apes and modern humans does the origin extend posteriorly to reach the ischial tuberosity. In all hominoids the muscle inserts onto the medial ridge of the linea aspera (or its equivalent); *Pongo* also has an insertion onto the intercondylar femoral line and in modern humans it may also insert onto the medial supracondylar femoral line. In *Hylobates* the muscle has been described as having more than two bellies, the 'long' and 'short' heads (Beddard, 1893; Hepburn, 1892), but in the hylobatids we dissected the muscle only has one belly. According to Sigmon (1974) in hylobatids and orangutans the adductor magnus also originates from the inferomedial border of the semitendinosus muscle and from the long head of the biceps femoris, but this was not the case in the specimens we dissected. According to Beddard (1893) in orangutans the short head of the adductor magnus is also inserted onto the lateral edge of the common tendon of insertion of the adductor longus; we observed this condition in one of the two orangutans we dissected. Innervation of the adductor magnus is by the obturator and sciatic nerves in modern humans. In the Asian apes and in *Pan* the short head of adductor magnus is innervated by the obturator nerve (Champneys, 1871; Hepburn, 1892; Sigmon, 1974) and the long head by the flexores femoris nerve (Sigmon, 1974), but in the specimens dissected by Champneys (1871) and Hepburn (1892) the long head was innervated by the sciatic nerve. In *Gorilla* the long head is innervated by the tibial nerve or by the flexores femoris nerve, while the short head is innervated by the obturator nerve (Raven, 1950; Gibbs, 1999). According to Gibbs (1999), the tibial nerve innervates the long head in half of gorillas.

According to Hepburn (1892) and Sigmon (1974), the *adductor minimus* is absent in at least one third of the apes; instead it has been described as a proximal division of the adductor magnus, a deep division of the adductor brevis (Hepburn, 1892), or as a distal division of the adductors (Appleton and Ghey, 1929). The adductor minimus is said to be absent in *Hylobates* (Sigmon, 1974), although its partial fusion with the adductor magnus can represent a state of differentiation towards an independent structure according to some authors (Hepburn, 1892). In our dissections we were only able to find the adductor minimus in *Gorilla*. It originated from the ischial ramus, and inferior and superior public ramus; when present in modern humans the muscle only originates from the inferior pubic ramus. In *Gorilla* the adductor minimus is innervated by the obturator nerve (Raven, 1950).

In all hominoids the *pectineus* originates from the superior pubic ramus; in orangutans its origin extends to the region of the pubic tubercle. In all hominoids it is inserted onto the lesser trochanter; that insertion extends to the pectineal line in modern humans and to the dorsal proximal femur in all apes. According to Appleton and Ghey (1929) the pectineus insertion is well-marked in *Gorilla* and *Hylobates*, but not in *Pan* and *Pongo*; this observation was corroborated by our dissections. Regarding *Pongo*, the literature refers to an origin from the superior pubic ramus (Boyer, 1935; Hepburn, 1892; Sigmon, 1974; Sonntag, 1924) and this was seen in one of our two orangutans; in the other orangutan the origin was from the pubic tubercle and just lateral to it (Ferrero, 2011). According to MacAlister (1871) the pectineus is bilaminar in some chimpanzees, but this was not the case in the chimpanzees dissected by us. Innervation of the pectineus is by the femoral nerve (L_2-L_4) in modern humans and in the apes (Champneys, 1871; Raven, 1950; Sigmon, 1974; Gibbs, 1999), although in *Pongo* and *Pan* it may also be by the anterior branch of the obturator nerve (Champneys, 1871; Boyer, 1935; Sigmon, 1974).

Contrary to the condition in other hominoids, in African apes the *gracilis* is a broad, flat muscle. In all hominoids the gracilis originates from the pubic body and its inferior ramus and it inserts onto the proximal part of the medial surface of the tibial diaphysis; in *Pongo* and *Pan* the insertion extends more anteriorly. According to Sonntag (1924), an additional insertion onto the superior fascia of the leg occurs in all apes. Hepburn (1892) described the origin of the gracilis in hylobatids extending across the body of the pubis, but this was not the case in our gibbons. In *Pongo* the insertion of the gracilis is said to be between the sartorius and the semitendinosus (Beddard, 1893; Boyer, 1935; Hepburn, 1892; Owen, 1830-1831; Sigmon, 1974; Sonntag, 1924); we found this condition in one orangutan, but in the other the insertion was distolateral to the semitendinosus (Ferrero, 2011). Innervation of the gracilis is by the obturator nerve (L_3-L_4) in modern humans and *Gorilla* (Raven, 1950), and specifically by a ventral branch of this nerve according to the literature review of Gibbs (1999) in the latter genus. In the Asian apes and in *Pan* it is by the anterior branch of the obturator nerve (Sigmon, 1974). The gracilis adducts, but also flexes and medially rotates the leg.

In all hominoids the *biceps femoris* originates from the ischial tuberosity (long head) and from the lateral ridge of the linea aspera (short head). In all hominoids there is an insertion onto the head of the fibula, in apes also onto the lateral condyle and tuberosity of the tibia, and in Asian apes also onto the capsule of the knee joint. In *Pongo* there are insertions onto the collateral fibular ligament and the lateral epicondyle of the femur.

The two heads of the biceps femoris are fused at their insertion in *Hylobates*, *Pan* and modern humans, but not in *Pongo* and *Gorilla*. In *Hylobates* we found a common origin of the long head with the ischiofemoralis, and not with the gluteus maximus as was claimed by Sigmon (1974). According to Beddard (1893) in *Pongo* the long head of the biceps femoris originates in common with the semimembranosus and quadratus femoris; we observed this in one of our two orangutans, in the other the long head was fused with the semimembranosus. Other authors describe a common origin with the gluteus maximus (Sigmon, 1974; Sonntag, 1924), but this condition was not found in our orangutans. Our gorilla had an insertion as found by other authors (see Gibbs, 1999), namely onto the lateral tibial condyle, tibial tuberosity and the head of the fibula.

In *Pan* the long head is said to have a common origin with the semimembranosus (Beddard, 1893; Champneys, 1872) or with the ischiofemoralis and gracilis (Champneys, 1872); in one of our chimpanzees the common origin was with the gluteus maximus. In one of the chimpanzees dissected by us, a belly of the long head merged with the insertion tendon of the short head, a feature not mentioned by any previous author.

The short head was described as absent in one specimen of *Pan* dissected by Sutton (1883), and some authors describe an additional insertion of the biceps femoris onto the lateral intermuscular septum of the thigh (Prejzner-Morawska and Urbanowick, 1971); we observed the latter in one of our chimpanzees. Innervation of the long head is by the flexor femoris nerve in apes and in modern humans; innervation by the tibial nerve was described by Raven (1950) and occurs in 1/2 of gorillas according to Gibbs (1999). The short head is innervated by the common fibular nerve, as is the case in other apes and in modern humans (Champneys, 1871; Raven, 1950; Sigmon, 1974; Gibbs, 1999).

In all hominoids the *semimembranosus* originates from the ischial tuberosity and inserts onto the medial condyle of the tibia and the tibial collateral ligament. In modern humans and *Gorilla* there is an additional posterior insertion onto the oblique popliteal ligament. In *Hylobates* the semimembranosus has been described as a long and flat, ribbon-like muscle (Sigmon, 1974) with a flat insertion tendon (Hepburn, 1892), but in our gibbon cadavers it was spindle-shaped with a more rounded tendon where it inserts. Sigmon (1974) describes the origin to be lateral to the origin of the semitendinosus muscle, but in our gibbons the origin of the semimembranosus is deep to that of the semitendinosus. In some chimpanzees the semimembranosus and semitendinosus are fused at their origin (Beddard, 1893; Champneys, 1872). Innervation of the semimembranosus is by the sciatic nerve (tibial portion L_5-S_2) in modern humans; in Asian apes and in *Pan* it is by the flexor femoris nerve (Sigmon, 1974). Innervation by the tibial nerve was described by Raven (1950) and according to the literature review of Gibbs (1999) it occurs in c.50% of gorillas.

The *semitendinosus* originates from the ischial tuberosity in all hominoids. In all but *Pan* it is inserted onto the proximal part of the medial surface of the tibial diaphysis; all apes also having an insertion onto the medial aspect of the tibial tuberosity, but this is not found in modern humans, and *Pongo* has an additional insertion onto the tibial collateral ligament. In some apes an aponeurotic expansion of the semitendinosus blends with the deep fascia of the leg (Boyer, 1935; Owen, 1830-1831; Sonntag, 1924). Innervation of the semitendinosus muscle is by the tibial nerve component of the sciatic nerve (L_5-S_2) in modern humans, and in Asian apes and *Pan* by the flexor femoris nerve (Sigmon, 1974). In *Gorilla* innervation by the tibial nerve was described by Raven (1950) and occurs in c.50% of gorillas according to the literature review of Gibbs (1999).

Table 3. Presence/absence of specific origins and insertions for the leg muscles in different hominoid taxa (see caption of Table 1)

Origin	*Hy*	*Pg*	*Gll*	*Pn*	*Hm*	Muscle	Insertion	*Hy*	*Pg*	*Gll*	*Pn*	*Hm*
Lateral tibial condyle	*	*	*	*	*	*Extensor digitorum longus*	Digits 2 to 5, dorsal aponeurosis	*	*	*	*	*
3/4 fibular proximal region					*		Digits 2 to 5, middle phalanx					*
Fibular head	*	*	*	*	*		Digits 2 to 5, distal phalanx			*		*
Fibular medial crest	*	*		*								
Fibular anterior surface			*									
Interosseous membrane	*	*	*	*	*							
Lateral tibial condyle	*			*		*Extensor hallucis longus*	Digit 1, dorsal aponeurosis		*		*	
Fibular medial surface	*	*	*	*	*		Digit 1, distal phalanx dorsal surface	*	*	*	*	*
Interosseous membrane		*	*	*	*		Digit 1, proximal phalanx dorsal surface			*		+
1/3 distal fibular anterior surface					*	*Fibularis tertius*	MV proximo-dorsal region					*
Interosseous membrane					*							
Fibular lateral surface	*	*		*	*	*Tibialis anterior*	Medial cuneiform, plantar surface	*	*		*	*
Lateral tibial condyle, distal region	*		*		*		Medial cuneiform, medial surface					*
Tibial anterior surface	*	*					Lateral cuneiform, plantar surface			*		

Origin	*Hy*	*Pg*	*Gll*	*Pn*	*Hm*	Muscle	Insertion	*Hy*	*Pg*	*Gll*	*Pn*	*Hm*
Tibial medial surface		*	*				MI proximal end	*	*	*	*	*
Tibial tuberosity	*											
Interosseous membrane		*	*	*	*							
Fibular distal anterior surface	*	*		*		*Fibularis brevis*	MV tuberosity	*	*	*	*	*
Fibular distal medial surface	*						Digit 5, proximal phalanx base	*		*	*	
Fibular distal lateral surface		*	*	*	*		Digit 5, medial phalanx base			*	#	
Fibular distal posterior surface		*					Extensor digitorum brevis tendon to digit 5		*	*		
							MV diaphysis				+	
Fibular head	*	*	*	*	*	*Fibularis longus*	MI tuberosity	*	*	*	*	*
Fibular proximal anterior surface	*						Fibular lateral malleolus			*		
Fibular proximal lateral surface	*	*	*		*		Intermediate cuneiform					*
Lateral tibial condyle	*			*								
Tibial collateral ligament	+											
Lateral femoral epicondyle	*	*	*	*	*	*Gastrocnemius*	Calcaneus posterior tuberosity	*	*	*	*	*
Medial femoral epicondyle	*	*	*	*	*							

Table 3. (Continued)

Origin	*Hy*	*Pg*	*Gll*	*Pn*	*Hm*	Muscle	Insertion	*Hy*	*Pg*	*Gll*	*Pn*	*Hm*
Knee joint capsule		*	*	*								
Fibular head, posterior surface	*	*	*	*	*	*Soleus*	Calcaneus posterior tuberosity	*	*	*	*	*
Fibular head, medial surface		*	*	*								
Fibular proximal posterior surface				*	*							
Tibial medial edge					*							
Lateral femoral epicondyle	+			*	+	*Plantaris*	Calcaneus posterior tuberosity	+	*		*	*
Femoral lateral supracondylar line					*							
Lateral tibial condyle, posterior region		*										
Oblique popliteal ligament					*							
Tibial diaphysis, posterior surface	*		*	*	*	*Flexor digitorum medialis*	Digit 2, distal phalanx base	*	*	*	*	*
Tibial diaphysis, lateral surface		*			+		Digit 3, distal phalanx base	*		*	*	*
Fibular diaphysis, medial surface		*			#		Digit 4, distal phalanx base	*	*	*	*	*

Origin	*Hy*	*Pg*	*Gll*	*Pn*	*Hm*	Muscle	Insertion	*Hy*	*Pg*	*Gll*	*Pn*	*Hm*
Interosseous membrane		*			+		Digit 5, distal phalanx base	*	*	*	*	*
Fibular head posterior surface	+	*	*	*		*Flexor digitorum lateralis*	Digit 1, distal phalanx base	*		*	*	*
Fibular proximal posterior surface	+	*	*	*	*		Digit 2, distal phalanx base	*		*	*	
Lateral femoral epicondyle		*					Digit 3, distal phalanx base	*	*	*	*	
Interosseous membrane	*	*	*	*	*		Digit 4, distal phalanx base	*	*	*	*	
Tibial collateral ligament		+					Digit 5, distal phalanx base			*		
Femoral lateral epicondyle	#	#	*		*	*Popliteus*	Tibial proximal posterior surface	*	*	*	*	*
Tibial lateral epicondyle	+			*	+		Tibial proximal lateral surface	+	*		*	+
Fibular head		*					Tibial proximal medial surface		+			
Popliteal oblique ligament	+			+	*		Tibial collateral ligament		*		*	+
Popliteal oblique ligament, sesamoid bone		*										
Knee joint capsule		*	*									
Lateral meniscus	+	+	+	+	*							
Fibular proximal medial surface	*	*	*	*	*	*Tibialis posterior*	Navicular base	*	*	*	*	*
Tibial proximal lateral surface	*	*	*	*	*		Cuboid tuberosity		*			*

Table 3. (Continued)

Origin	*Hy*	*Pg*	*Gll*	*Pn*	*Hm*	Muscle	Insertion	*Hy*	*Pg*	*Gll*	*Pn*	*Hm*
Interosseous membrane	*	*	*	*	*	*Tibialis posterior* (cont.)	Medial cuneiform		#			*
							Intermediate cuneiform		*		*	*
							Lateral cuneiform		*	*	*	*
							MII to MIV, base of plantar surface	*			*	*
							Lateral malleolus			*		
							Plantar ligaments	*	*		*	
							Long fibular muscle sheath	*	*	*	*	

Muscles of the Leg (Table 3)

The leg muscles are divided into three groups: anterior (extensor digitorum longus, extensor hallucis longus, fibularis tertius and tibialis anterior), posterior (gastrocnemius, soleus, plantaris, flexor digitorum medialis, flexor digitorum lateralis, popliteus and tibialis posterior) and lateral (fibularis brevis and fibularis longus) that are separated by the interosseous membrane and the two intermuscular septa of the leg.

In all hominoids the *extensor digitorum longus* originates from the lateral tibial condyle, the head of the fibula and the adjacent interosseous membrane; in gorillas there is also an origin from the anterior surface of the shaft of the fibula. The insertion in all taxa is onto the dorsal aponeurosis of digits 2-5; but in one of the three hylobatids dissected by us the tendons of this muscle are not clearly defined and separated as is usually the case in hominoids. Innervation of the muscle is by the tibial component of the sciatic nerve (L_5-S_1) in modern humans, and by branches of the deep fibular nerve in *Pan paniscus* and *Gorilla* (Raven, 1950; Miller, 1952; Gibbs, 1999).

In all hominoids the *extensor hallucis longus* originates from the medial aspect of the fibula shaft and in *Hylobates* and *Pan* there is an additional origin from the lateral condyle of the tibia. In all taxa insertion is onto the dorsum of the distal phalanx of the hallux, although our gorilla specimen has an additional insertion onto the dorsum of the proximal phalanx of this digit. An origin from the interosseous membrane of the leg was also reported for this taxon (Beddard, 1893; Boyer, 1935; Lewis, 1966), but we only found this condition in one orangutan. Innervation of the muscle is by the tibial component of the sciatic nerve (L_5-S_1) in modern humans, and by the deep fibular nerve in *Pan paniscus* and *Gorilla* (Raven, 1950; Miller, 1952; Gibbs, 1999).

According to the literature review of Gibbs (1999), the *fibularis tertius* is present in 50% of *Hylobates*, 30% of *Gorilla*, 5% of *Pan* 5% and 95% of modern humans; it is absent in *Pongo*. When present the muscle origin is associated with the extensor digitorum longus and its insertion is onto MV; it weakly everts and dorsiflexes the foot. It is innervated by the deep fibular nerve in modern humans. We could not find a fibularis tertius in any of the apes dissected by us.

In all hominoids the *tibialis anterior* has a partial origin from the lateral aspect of the proximal tibia. In *Gorilla* and *Pongo* the origin is from the medial aspect of the proximal tibia, and in *Pongo* and *Hylobates* from the anterior aspect of the proximal tibia. According to Boyer (1935) in *Pongo* the origin extends to the anterior and posterior aspects of the proximal tibia; we found this to be the case in one of our two orangutans. In all of the apes dissected by us the muscle divides into two bellies close to the terminal third of its length; in modern humans such a division is described as a variation (Warwick and Williams, 1973). In all hominoids the muscle inserts onto the proximal portion of MI and onto the plantar surface of the medial cuneiform (Beddard, 1893; Champneys, 1872; Hepburn, 1892; Humphry, 1866-1867; Lewis, 1966; MacDowell, 1910; Miller, 1952; Sutton, 1883); in modern humans it extends onto the medial aspect of the medial cuneiform (Moore and Dalle, 2002). In *Gorilla* there is an insertion onto the plantar surface of the lateral cuneiform. In the hylobatids dissected by us there is an origin from the tibial tuberosity. Innervation of the tibialis anterior is by the deep fibular nerve (L_4-S_1) in modern humans and *Gorilla* (Raven, 1950; Gibbs, 1999), and by a branch of the deep fibular nerve in *Pan paniscus* (Miller, 1952). The muscle dorsiflexes and inverts the foot.

Both the *fibularis brevis* and the fibularis longus are evertors of the foot in modern humans. The origin of the fibularis brevis is from the lateral aspect of the distal fibula in all hominoids, except for *Hylobates*, in which the muscle originates from the anterior and medial aspect of this bone. In *Pongo* there is an additional origin from the posterior surface of the distal fibula. The insertion in all hominoids is onto the tuberosity of MV, extending to the proximal phalanx of digit 5 in all apes except *Pongo*, and to the base of the middle phalanx of this digit in *Gorilla* only. Our dissections confirm the insertion of the muscle described in the literature (Beddard, 1893; Hepburn, 1892; Lewis, 1966; MacDowell, 1910), but in one of our three hylobatids there is a thin additional tendon inserting laterally onto the dorsal aponeurosis of digit 5. Some authors describe an insertion onto the extensor digitorum brevis tendon in *Pongo* (Beddard, 1893; Hepburn, 1892; MacDowell, 1910; Sonntag, 1924); we confirmed this in one of our orangutan cadavers. Hepburn (1892) describes some specimens of *Pongo* with a third insertion, which according to Boyer (1935) is associated with connective tissue. In one of our orangutans there is effectively a third belly of the fibularis brevis giving rise to a thin tendon that is inserted onto the plantar fascia. Some authors have found a thin tendon of the muscle extending onto the proximal and middle phalanges of digit 5 in *Gorilla*, as occasionally found in modern humans (MacAlister, 1873; Raven, 1590; Gibbs, 1999). An insertion onto the proximal and middle phalanges of digit 5 has also been reported in *Pan* (Champneys, 1872; MacDowell, 1910; Prejzner-Morawska and Urbanowick, 1971; Vrolik, 1841); in one of our chimpanzees there was an insertion onto the base of the proximal phalanx of digit 5, while in another chimpanzee there was an insertion onto the MV shaft, not onto digit 5. Innervation of the muscle is by the superficial fibular nerve (L_4-S_1) in modern humans and *Gorilla* (Raven, 1950; Gibbs, 1999) and by branches of the superficial fibular nerve in *Pan paniscus* (Miller, 1952).

The *fibularis longus* originates from the head of the fibula in all hominoids, and the attachment extends to the lateral aspect of the proximal fibula shaft in all taxa except *Pan*. In *Hylobates* there is also an origin from the anterior aspect of the proximal fibula and from the tibial collateral ligament, which was not described by previous authors. The insertion of this muscle in all the hominoids is onto the tuberosity of MI (Manners-Smith, 1908; Miller, 1952; Oxnard and Lisowski, 1980), plus in modern humans onto the intermediate cuneiform bone. According to Lewis (1966) there is a fibrous insertion onto MV; we found this condition in one chimpanzee. Innervation is by the superficial fibular nerve (L_4-S_1) in modern humans and according to Miller (1952) by branches of the superficial fibular nerve in *Pan paniscus* and *Gorilla*.

The superficial group of *posterior muscles* of the leg consists of gastrocnemius, soleus and plantaris. The first two are often called the 'triceps surae' or 'calf' muscles, and some authors consider that the plantaris is a vestigial 'surae' muscle. The muscles of the calf of modern humans are twice as heavy as those of *Pan* relative to total body weight (Zihlman and Brunk, 1979). The development of the triceps surae in modern humans reflects the fact that these muscles are related to the movements of the foot during the second half of the stance phase of bipedal locomotion. The well-developed Achilles tendon in modern humans is also an indication of the importance of these muscles for bipedal locomotion; according to Alexander (1984) the Achilles tendon acts like a spring, allowing modern humans to retain a considerable amount of energy, particularly during running. The relatively small Achilles tendon of apes suggests a reduced efficiency of this tendon as an energy store in these hominoids (Aiello and Dean, 1990). This observation seems to be supported by the fact that in hylobatids, which exhibit rapid and frequent bipedal movements, the tendon of the gastrocnemius is more developed than in other apes (Ferrero, 2011).

In all hominoids the *gastrocnemius* originates from the lateral and medial epicondyles of the femur; in *Pongo*, *Gorilla* and *Pan* the origins encroach onto the adjacent capsule of the knee joint. Insertion in all hominoids is onto the posterior tuberosity of the calcaneus. Some authors described sesamoid bones in each of the heads of origin of the muscle in *Hylobates* (Frey, 1913;

Hepburn, 1892), but not in the other apes (Champneys, 1871; Frey, 1913; Owen, 1830-1831; Raven, 1950; Prejzner-Morawska and Urbanowick, 1972); we found no sesamoid bones in any of the apes we dissected. Innervation of the muscle is by the tibial nerve (S_1-S_2) in modern humans and *Gorilla* (Raven, 1950; Gibbs, 1999) and in *Pan paniscus*, according to Miller (1952), by branches of the tibial nerve. The muscle is a plantar flexor of the foot and a flexor of the leg.

The *soleus* of *Gorilla* and *Pan* is more bulky than that of the Asian apes. In all hominoids the soleus has an origin from the posterior aspect of the head of the fibula; modern humans have a second origin from a bony ridge on the medial surface of the tibia. Insertion in all hominoids is from the posterior tuberosity of the calcaneus. In 3% of modern humans there is an accessory soleus muscle distal and medial to the Achilles tendon (Moore and Dalley, 2002); we found such an accessory muscle in two specimens of *Pongo* dissected by us (Figure. 1). Innervation of the soleus is by the tibial component of the sciatic nerve (S_1-S_2) in modern humans and *Gorilla* (Raven, 1950; Gibbs, 1999) and by branches of the tibial nerve in *Pan paniscus* (Miller, 1952). In all hominoids the soleus plantar flexes the ankle joint.

The *plantaris* is reported as present in 90-95% of modern humans, in some orangutans and chimpanzees, but absent in gorillas and hylobatids (Gibbs, 1999), nevertheless in the three hylobatids dissected by us the muscle is present (Table 3, Figure. 2) (Ferrero, 2011). The muscle was reported in two orangutans by previous authors (Primorse, 1899; Sandifort, 1840); it was present in one of our two orangutans. It was described as absent in three-fifths of chimpanzees (Chapman, 1879; Embleton, 1864; Hepburn, 1892; Loth 1908; Loth, 1914; Michaëlis, 1903; Rozycki, 1922; Traill, 1817; Prejzner-Morawska and Urbanowick, 1972), unilaterally present in one fifth of them (Champneys, 1872; MacAlister, 1871; Wilder, 1862), and bilaterally in two fifths (Beddard, 1893; Duckworth, 1898; Dwight, 1895; Frey, 1913; Hartmann, 1883; Humphry, 1866-1867; Huxley, 1864; MacDowell, 1910; Miller, 1952; Sutton, 1883; Prejzner-Morawska and Urbanowick, 1972; Vrolik, 1841). We found it unilaterally in two chimpanzees. Only *Hylobates* and *Pan* have an origin of the plantaris from the lateral femoral epicondyle. In *Pongo* it is from the posterior aspect of the lateral tibial condyle, and in modern humans from the lateral supracondylar line of the femur and the oblique popliteal ligament. It is always inserted onto the posterior tuberosity of the calcaneus. Innervation is by the tibial nerve (L_4-S_1) in modern humans and by branches of the tibial nerve in *Pan paniscus* (Miller, 1952). The plantaris is better-developed in non-anthropoid primates than in hominoids (Farslow and Sigmon, 1986). In

Hylobates, *Pan* and modern humans plantaris flexes the leg and plantar flexes the foot, but in *Pongo* it acts only as a plantar flexor of the foot because its origin is restricted to the tibia.

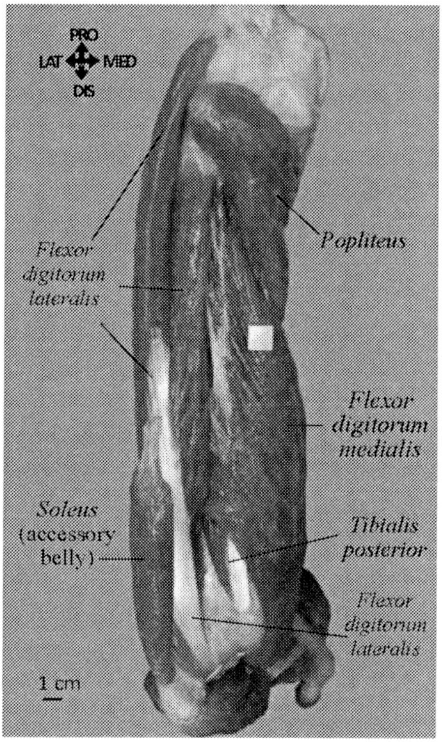

Figure 1. VU PP1 (*Pongo pygmaeus*, adult female), dorsal view of the left leg showing an accessory belly of soleus. In this and the other figures, pro, lat, med and dis mean proximal, lateral, medial and distal, respectively.

The flexor hallucis longus is usually completely separate from the flexor digitorum longus in modern humans, although sometimes these two muscles may be blended (Aiello and Dean, 1990). In one of the gibbons dissected by us there is a merging between the *flexor digitorum medialis* (named flexor digitorum longus) and the flexor digitorum lateralis (named flexor hallucis longus) (Ferrero, 2011). In all hominoids except *Pongo* the flexor digitorum medialis originates from the posterior surface of the tibial diaphysis, in modern humans and *Pongo* also from its lateral aspect, and also from the medial surface of the fibular diaphysis (although this latter origin was not observed on the modern human dissected by us).

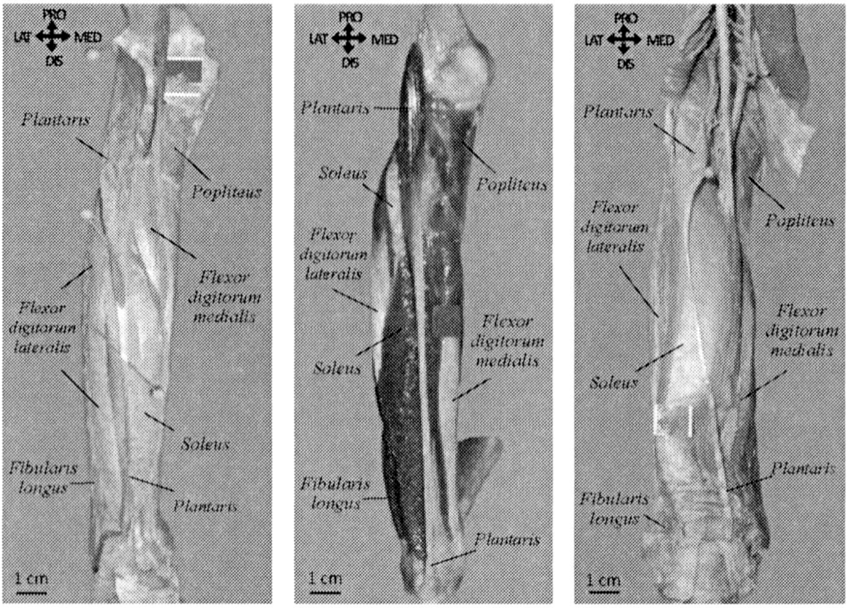

Figure 2. From left to the right, GWU HL1 (*Hylobates lar*, juvenile female), VU HG1 (*H. (Nomascus) gabriellae*, adult male) and HU HL1 (*H. lar*, adult male), showing the dorsal aspect of the leg; the plantaris is present in the three specimens.

In one of our chimpanzees there was an additional origin from the fascia associated with the tibialis posterior. Digit 5 receives a tendon in all hominoids (Beddard, 1893; Duvernoy, 1853; Hepburn, 1892; Keith, 1894; Lewis, 1962; MacDowell, 1910; Ferrero, 2011). In all of the apes and modern human studied there are tendons to digits 4 and 2, and in all except orangutans there is an insertion onto digit 3. In *Hylobates* there is usually an insertion onto digit 5 (Beddard, 1893; Duvernoy, 1853; Hepburn, 1892; Keith , 1894; Lewis, 1962; MacDowell, 1910), and digit 4 (Hepburn, 1892; Lewis, 1962), while digit 3 receives a tendon in one-third of the cases (Hepburn, 1892), digit 2 in two-third of the cases (Lewis, 1962); the tendon to this latter digit was fused with the flexor digitorum lateralis in one specimen (Hepburn, 1892). In the three hylobatids dissected by us there was a different configuration for each specimen: one of them presented a tendon of insertion going to the distal phalanx of digits 2, 4 and 5, in another specimen to the distal phalanx of digits 4 and 5, and the third specimen presented an unusual configuration with insertions onto the distal phalanx of digits 2 to 5. Innervation is by the deep fibular nerve (L_4-S_1) in modern humans, by branches of the tibial nerve in *Pan paniscus* (Miller, 1952), and by the tibial nerve in *Gorilla* (Raven, 1950;

Gibbs, 1999). The flexor digitorum medialis flexes the pedal digits and plantar flexes the foot.

In all hominoids the *flexor digitorum lateralis* originates from the proximal part of the posterior aspect of the fibula and the interosseous membrane; in all apes the attachment extends to the fibular head and in *Pongo* to the lateral femoral epicondyle. Beddard (1893) and MacDowell (1910) report an origin from the interosseous membrane and the lateral intermuscular septum in hylobatids, but do not describe an origin from the posterior aspect of the proximal diaphysis of the fibula and from the posterior aspect of its head, a condition we observed in our gibbons. In one of our gibbons there were two heads of origin, one from the posterior aspect of the fibula and the other from the interosseous membrane, proximal to the origin of tibialis posterior. For *Pongo* there are reports of an origin from the lateral femoral condyle (Beddard, 1893; Boyer, 1935; Hepburn, 1892, Owen, 1830-1831; Sonntag, 1924); this was the case in our specimens, but in addition we also found an origin from the tibial collateral ligament (Ferrero, 2011). *Pongo* lacks an insertion onto the hallux (Straus, 1942), but other apes have an insertion onto the hallux, and all hominoids except modern humans have an insertion onto digits 3 and 4; digit 2 only receives a tendon in African apes and hylobatids, and digit 5 only receives a tendon in *Gorilla*. The insertion of the flexor digitorum lateralis is onto the base of the terminal phalanx of the hallux in all apes (Beddard, 1893, Hepburn, 1892, Keith, 1894; Lewis, 1962; MacDowell, 1910; Michaëlis, 1903), with the exception, as stated above, of *Pongo* (Boyer, 1935; Brooks, 1887; Chapman, 1880; Hepburn, 1892; Owen, 1830-1831; Primrose, 1900, Ferrero, 2011). According to Owen (1830-1831) in *Pongo* the space that is normally occupied by the flexor digitorum lateralis is occupied by a small muscle running from MI to the phalanx of the hallux; we did not found this condition in our orangutans. Innervation of the flexor digitorum lateralis is by the deep fibular nerve (L_4-S_1) in modern humans, by branches of the tibial nerve in *Pan paniscus* (Miller, 1952), and by the tibial nerve in *Gorilla* (Raven, 1950; Gibbs, 1999). The muscle is a flexor of the digits to which it is inserted and a plantar flexor of the foot.

In all hominoids the *popliteus* originates from the lateral meniscus (this origin is described here for the first time, for apes). In the hylobatids dissected by us there was no origin from the lateral femoral epicondyle as described by other authors, nor was there a sesamoid bone in the tendon of origin of the muscle as reported in one gibbon by Pearson and Davin (1921). In our gibbons there is instead an origin from the lateral tibial epicondyle, from the oblique popliteal ligament and from the lateral meniscus (Ferrero, 2011), but not from the femur, and an insertion onto the posterior aspect of the tibial shaft as described by Beddard (1893) as well as onto the lateral aspect of the tibial shaft. Some authors reported a head of origin from the lateral femoral epicondyle in *Pongo* (Beddard, 1893; Boyer, 1935; Hepburn, 1892; Sonntag, 1924), but this configuration is not found in our orangutans; according to Hepburn (1892) there is a second head of origin from the fibular head and adjacent to the knee joint capsule, this configuration is found in one of our orangutans. In the other orangutan there is a sesamoid bone in the origin from popliteal oblique ligament, as reported in eighteen of the twenty *Pongo* described by Boyer (1935), Forster (1903), Fürst (1903), Le Minor (1992), Owen (1830-1831), Pearson and Davin (1921), Vallois (1914) and van Westrienen (1907). Our orangutans have an insertion onto the posterior and lateral aspect of the tibia as described by other authors (Beddard, 1893; Boyer, 1935; Sonntag, 1924; Ferrero, 2011), but also onto the medial aspect of this bone and onto the collateral tibial ligament as in *Pan* and modern human. *Gorilla* and modern humans were the ones who presented an origin from the femoral lateral epicondyle, which has an important implication in locomotion as will be discussed below In *Pan* we observed an additional origin from the oblique popliteal ligament not described by previous authors. Innervation of the popliteus is by the tibial branch of the sciatic nerve (S_1-S_2) in modern humans, by a branch of the tibial nerve (Hepburn, 1892) in Asian apes and *Pan*, and by the tibial nerve in gorillas (Raven, 1950; Gibbs, 1999). The muscle flexes and laterally rotates the thigh with a fixed leg or medially rotates the leg with a fixed thigh in modern humans, thus "unlocking" the fully extended knee (Aiello and Dean, 1990). The popliteus of the modern human specimen dissected by us does not have an origin from the lateral tibial epicondyle (Table 3; Figure. 3). For some authors, in apes, the muscle promotes medial rotation of the tibia in a knee joint that is more mobile than in modern humans, and it may also play an important role in helping to stabilize the knee joint (Aiello and Dean, 1990). Based on the results obtained in this study, the only ape in which the muscle can act as a powerful medial rotator of the tibia is the *Gorilla*, because this is the only taxon in which the popliteus has an origin from the lateral femoral epicondyle (Ferrero, 2011). We plan to undertake further detailed studies on the detailed comparative anatomy and evolution of this muscle.

The *tibialis posterior* originates from the medial and lateral aspects of the proximal and middle parts of the fibula and tibial shaft respectively, and from the interosseous membrane in all hominoids. In all taxa there is an insertion onto the navicular, and in modern humans also onto the medial cuneiform. The insertion of the tibialis posterior in modern humans differs from that of *Pan* because in the former it is blended with the tendon of the flexor hallucis brevis. Lewis (1964b) argued that the blending is an adaptation to bipedal locomotion and is related to the stabilization of the longitudinal arch of the modern human foot. Regarding *Pongo*, in our orangutans instead of insertion onto all three cuneiforms (Boyer, 1935; Lewis, 1964a), we found an insertion onto the intermediate and lateral cuneiforms. Innervation of the muscle is by the tibial branch of the sciatic nerve (S_1-S_2) in modern humans, by branches of the tibial nerve in *Pan paniscus* (Miller, 1952), and by the tibial nerve in *Gorilla* (Raven, 1950; Gibbs, 1999). The tibialis posterior plantar flexes and inverts the foot during arboreal locomotion in *Pan* (Lewis, 1980).

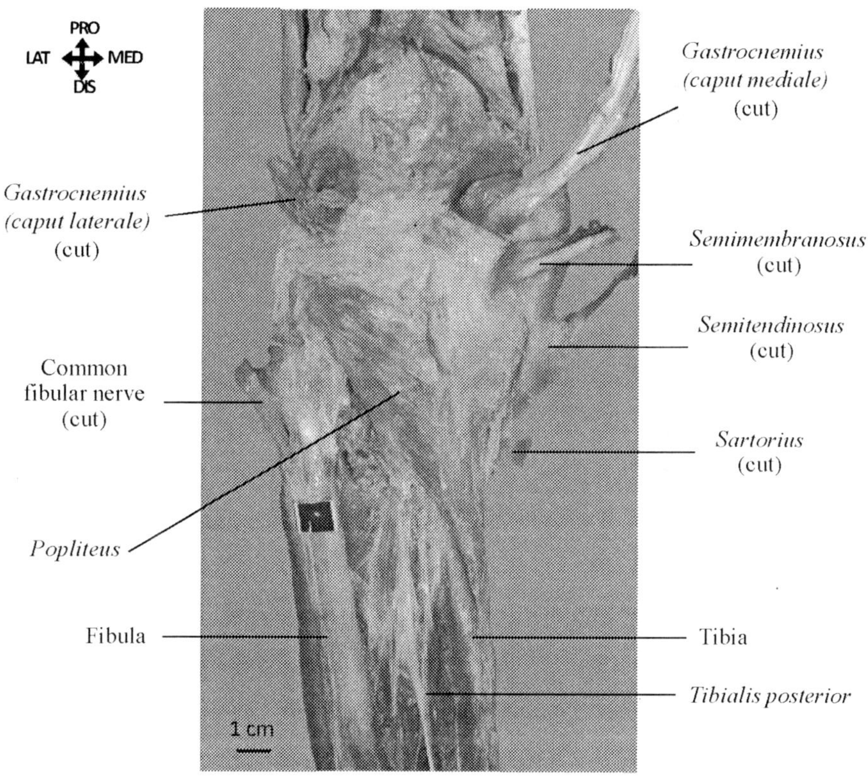

Figure 3. HU HS1E (*Homo sapiens*, adult female), posterior view of the popliteal region of the right leg. Note that the popliteus originates from the lateral epicondyle of the tibia and from the popliteal oblique ligament.

Muscles of the Foot (Table 4)

The muscles of the foot are divided into two groups, those on the dorsum and those in the sole of the foot. In all hominoids the *extensor digitorum brevis* originates from the lateral surface of the calcaneus, and in all except for *Gorilla* also from its dorsal surface. The usual insertion in hominoids is onto digits 2, 3 and 4 (Aiello and Dean, 1990); in modern humans it occasionally inserts onto the hallux. In half of the hylobatids studied by Lewis (1966) there was an additional insertion onto digit 5 that merged with the fibularis brevis; this condition was not found in our gibbons. An insertion onto digit 5, not observed in our specimens, has also been described in *Pan* (Lewis, 1966; Miller, 1952; Warwick and Williams, 1973). Innervation of the extensor digitorum brevis is by the deep fibular nerve (L_3-S_1) in modern humans, by branches of the fibular nerve in *Pan paniscus* (Miller, 1952) and by the deep fibular nerve in *Gorilla* (Raven, 1950; Gibbs, 1999).

The origin of the *extensor hallucis brevis* in all hominoids is from the medial surface of the calcaneus, and in all but *Gorilla* also from its dorsal surface; in *Pongo* there is also an origin from the navicular. Insertion in all hominoids is onto the proximal phalanx of the hallux. It is claimed that in Asian apes the muscle arises from the extensor digitorum brevis (Beddard, 1893; Boyer, 1935; Hepburn, 1892; Lewis, 1966; Sonntag, 1924), and according to Boyer (1935) the two muscles are fused, but in our specimens this was not the case. In a single specimen of *Pan* reported by Dwight (1895) there was an insertion onto the medial phalanx of the hallux. Innervation of the muscle is by the deep fibular nerve (L_3-S_1) in modern humans, by branches of the fibular nerve in *Pan paniscus* (Miller, 1952) and by the deep fibular nerve in *Gorilla* (Raven, 1950; Gibbs, 1999).

In all hominoids the *abductor digiti minimi* originates from the plantar fascia and the lateral aspects of the calcaneus, extending to the medial plantar aspect in all taxa except *Pan*. In all taxa insertion is onto the proximal phalanx of digit 5 with an extension to the base of MV. This muscle has not been described previously in hylobatids, but it was present in the hylobatids dissected by us (Table 4). In *Hylobates* some of the muscular fibers insert onto the base of MV, but the main insertion is onto the lateral aspect of the proximal phalanx of digit 5. According to Sonntag (1924) in *Pongo* there is an origin from the fibularis longus sheath; we found this condition in one of our orangutans. In *Gorilla* we observed the configuration described by Gibbs (1999) but also found an additional origin from the plantar aponeurosis.

Table 4. Presence/absence of specific origins and insertions for the extrinsic and intrinsic muscles of the foot in different hominoid taxa (see caption of Table 1)

Origin	*Hy*	*Pg*	*Gll*	*Pn*	*Hm*	Muscle	Insertion	*Hy*	*Pg*	*Gll*	*Pn*	*Hm*
Calcaneus, lateral surface	*	*	*	*	*	*Extensor digitorum brevis*	Digit 1 dorsal aponeurosis lateral surface					*
Calcaneus, dorsal surface	*	*		*	*		Digit 2 dorsal aponeurosis lateral surface	*	*	*	*	*
							Digit 3 dorsal aponeurosis lateral surface	*	*	*	*	*
							Digit 4 dorsal aponeurosis lateral surface	*	*	*	*	*
Calcaneus, medial surface	*	*	*	*	*	*Extensor hallucis brevis*	Digit 1 proximal phalanx dorsal	*	*	*	*	*
Calcaneus, dorsal surface	*	*		*	*		Digit 1 medial phalanx dorsal				#	
Navicular		*										
Calcaneus, medial plantar aspect	+	*	*		*	*Abductor digiti minimi*	MV base	+	*	*	*	+
Calcaneus, lateral plantar aspect	+	*	*	*	*		Digit 5proximal phalanx lateral region	+	*	*	*	*
Plantar fascia	+	*	*	*	*		Fibular longus and brevis muscle sheath					+
Fibular longus muscle sheath		*										
Calcaneus, medial plantar aspect	*	*	*	*	*	*Abductor hallucis*	Digit 1 proximal phalanx medial region	*	*	*	*	*
Calcaneus, posterior aspect		*					MI lateral region			*		

Table 4. (Continued)

Origin	*Hy*	*Pg*	*Gll*	*Pn*	*Hm*	Muscle	Insertion	*Hy*	*Pg*	*Gll*	*Pn*	*Hm*
Cuboid			*			*Abductor hallucis* (cont.)	Intermediate cuneiform				#	
Plantar fascia	*	*	*	*	*		Lateral cuneiform				#	
Flexor retinaculum					*							
MV tuberosity lateral aspect	+	*			*	*Abductor metatarsi quinti*	Digit 5 proximal phalanx lateral region	+	+			*
Calcaneus, lateral aspect			*				MV tuberosity		#	*		
Calcaneus, plantar aspect		#										
Calcaneus, medial plantar aspect	*	*	*	*	*	*Flexor digitorum brevis*	Digit 2 medial phalanx lateral and medial region	*	*	*	*	*
Calcaneus, anterior plantar aspect		*	*	*			Digit 3 medial phalanx lateral and medial region	*	*	*	*	*
Calcaneus, posterior plantar aspect		*					Digit 4 medial phalanx lateral and medial region	*	*	*	*	*
Plantar fascia	*	*	+	*	*		Digit 5 medial phalanx lateral and medial region	*				#
Flexor digitorum medialis tendon			*	*								
Calcaneal tendon		#										
Soleus medial surface		#										
Calcaneus, lateral plantar surface			*	*	*	*Quadratus plantae*	Flexor digitorum medialis tendon to digit 5 laterally			*	*	*
Calcaneus, medial plantar surface					*							

Origin	*Hy*	*Pg*	*Gll*	*Pn*	*Hm*	Muscle	Insertion	*Hy*	*Pg*	*Gll*	*Pn*	*Hm*
Flexor digitorum medialis tendons	*	*	*	*	*	*Lumbricales*	Digits 2 to 5 proximal phalanx base lateral region	*	*	*	*	*
Flexor digitorum lateralis tendons	*	*	*	*								
MII plantar surface	*	*	*	*	*	*Adductor hallucis*	MI base	*	*	*		
MIII plantar surface	*	*	*	*	*		MI head	*	*	*	*	
MIV plantar surface	*		*	*	#		Digit 1 proximal phalanx	*	*	*	*	*
Plantar interosseous fascia		*	*	*			Metatarsophalangeal joint of digit 1			*		
Fibularis longus tendons		*		*			Interphalangeal joint of digit			*		
Metatarsophalangeal joints	*	*		*	#							
Cuboid				*								
MIV base			*			*Flexor digiti minimi brevis*	Digit 5, proximal phalanx lateral region	+		*	*	
MV base	+	*	*	*	*		Digit 5, proximal phalanx plantar region	+		*	*	*
Fibular longus muscle sheath		*					Digit 5, proximal phalanx medial region	+	*			
Plantar tarsometatarsal ligament	+	+	*	*								
Plantar fascia		*										
MI base	*		*	*		*Flexor hallucis brevis*	Digit 1, proximal phalanx lateral region	*	*	*	*	*

Table 4. (Continued)

Origin	*Hy*	*Pg*	*Gll*	*Pn*	*Hm*	Muscle	Insertion	*Hy*	*Pg*	*Gll*	*Pn*	*Hm*
Calcaneus			*			*Flexor hallucis brevis* (cont.)	Digit 1, proximal phalanx medial region	*	*	*	*	*
Lateral cuneiform		*		#	*		MI distal region	*	*	*	*	
Intermediate cuneiform		*	*	*			Metatarsophalangeal joint of digit 1			*		
Medial cuneiform	*	*		*								
Navicular		*	*									
Cuboid				#	*							
First tarsometatarsal joint			*									
Cuneonavicular plantar ligament	*			*								
Tibialis posterior tendon		*		*								
MV base			*	*		*Opponens digiti minimi*	Digit 4, proximal phalanx lateral region		*		*	*
MV distal half					*		MV diaphysis, lateral edge		*	*	*	
Fibular longus muscle sheath		*										
Plantar fascia		*										
Plantar interosseous ligaments				*								
Medial cuneiform		*	*			*Interossei dorsales*	Digit 2, proximal phalanx medial region	*	*	*	*	*

Origin	*Hy*	*Pg*	*Gll*	*Pn*	*Hm*	Muscle	Insertion	*Hy*	*Pg*	*Gll*	*Pn*	*Hm*
MI lateral surface	*	*		*	*	*Interossei dorsales* (cont.)	Digit 3, proximal phalanx medial region		*	*	*	
MII lateral and medial surface	*	*	*	*	*		Digit 2, proximal phalanx lateral region	*				*
MIII lateral and medial surface	*	*	*	*	*		Digit 3, proximal phalanx lateral region	*	*	*	*	*
MIV lateral and medial surface	*	*	*	*	*		Digit 4, proximal phalanx lateral region	*	*	*	*	*
MV medial surface	*	*	*	*	*							
Navicular		*				*Interossei plantares*	Digit 2, proximal phalanx medial region	+				
MI medial surface		*					Digit 3, proximal phalanx medial region					*
MII medial surface	+	*	*	*			Digit 4, proximal phalanx medial region	+		*	*	*
MIII medial surface			*	*	*		Digit 5, proximal phalanx medial region	+		*	*	*
MIV medial surface	+	*	*		*		Digit 2, proximal phalanx lateral region		*	*	*	
MV medial surface	+	*	*				Digit 4, proximal phalanx lateral region		*			
MIII lateral surface		*					Digit 5, proximal phalanx lateral region		*			
MIV lateral surface				*								

In *Pan* we observed the origin often described in the literature (Beddard, 1893; Champneys, 1872; Latimer and Lovejoy, 1989; Miller, 1952). Innervation of the muscle is by the lateral plantar nerve (L_4-L_5) in modern humans and in *Pan paniscus* and *Gorilla* (Raven, 1950; Miller, 1952; Gibbs, 1999). One remarkable characteristic of the modern human foot is the relatively large size of the abductor muscles, particularly of the abductor hallucis.

One would expect that the abduction of the hallux would not be important in modern humans in which powerful opposition of the hallux has become lost, but electromyographic analyses confirmed the importance of the abductors of digits 1 and 5 in the stabilization of the foot during bipedal locomotion (Mann and Inman, 1964; Gray and Basmajian, 1968).

The *abductor hallucis* originates in all hominoids from the plantar fascia and the medial and plantar aspects of the calcaneus, extending to the posterior surface of this bone in *Pongo* and to the cuboid in *Gorilla.* In all taxa, it is inserted onto the proximal phalanx of digit 1, spreading to the lateral aspect of MI in *Gorilla.* No sesamoid bones were found in the tendon of insertion of this muscle in hylobatids, nor did we observe the partial fusion of the abductor hallucis and the flexor hallucis brevis muscles as described by other authors (Beddard, 1893; Bischoff, 1870; Brooks, 1887; Ruge, 1878b). Hylobatids occasionally have a muscle belly that is independent of the abductor hallucis located on the base of MI and some researchers have given it its own name 'adductor ossis metacarpi hallucis' (Brooks, 1887; Hepburn, 1892); we found this belly in one of the gibbons dissected by us. A comparable structure has also been described in *Pongo* (Brooks, 1887; Church, 1861-1862; Hepburn, 1892; Sonntag, 1924), but it was not found in the orangutans. A similar structure was reported in *Gorilla* by Gibbs (1999). In *Pan*, Vrolik (1841) described an additional insertion of the abductor hallucis onto the middle and lateral cuneiforms, nevertheless these insertions were not found in any of the chimpanzees dissected by us, but we did find an independent 'abductor ossis metacarpi hallucis' in one of our chimpanzees. Innervation of the abductor hallucis is by the medial plantar nerve (L_4-L_5) in modern humans and in *Pan paniscus* and *Gorilla* (Raven, 1950; Miller, 1952; Gibbs, 1999).

The *abductor metatarsi quinti* has been described in apes as separate from the abductor digiti minimi brevis, and as a variant in modern humans (Gibbs, 1999). It has been referred to as the 'abductor ossis metacarpi digiti quinti' (Raven, 1950) and the 'abductor os metatarsi digiti minimi' (Gibbs, 1999). This muscle has not been described previously in hylobatids, but it was present in the hylobatids dissected by us (Table 4). An origin from the lateral aspect of the tuberosity of MV is found in *Hylobates*, *Pongo* and modern humans. According to Primrose (1899) in orangutans the muscle takes origin from the plantar surface of the calcaneus, but in the orangutans dissected by us such an origin was not observed.

In *Gorilla* it has an origin from the lateral side of the calcaneus. Insertion is onto the proximal phalanx of digit 5 in *Hylobates*, *Pongo* and modern humans, and onto the tuberosity of MV in *Gorilla*. . However in one of the orangutans dissected by us there was also an insertion onto the proximal phalanx of digit 5. This muscle was found in two of the three hylobatids and the gorilla dissected by us, but not in the chimpanzees. Innervation of the muscle is by the lateral plantar nerve (Raven, 1950; Gibbs, 1999) in *Gorilla*, its action being mainly associated with the abduction of the fifth metatarsal.

In hominoids the origin of the *flexor digitorum brevis* is from the medial plantar aspect of the calcaneus, extending to the anterior aspect of this bone in *Pongo*, *Gorilla* and *Pan*, and even to the posterior plantar aspect of the bone in *Pongo*. African apes also have an origin from the tendon of the flexor digitorum medialis.

The insertion in all hominoids is onto the middle phalanges of digits 2, 3 and 4, extending to digit 5 in *Hylobates* and in modern humans (according to the literature, because such an insertion was not found in the modern human dissected by us).

In one of our hylobatids there is an unusual insertion onto digit 2; another hylobatid has insertions onto digits 2, 3 and 4, and in the third it is inserted onto digits 2 to 5 (Figure. 4). Boyer (1935) reported an origin from the calcaneus and from the medial aspect of the soleus in *Pongo*; however these origins were not observed in the orangutans dissected. In *Gorilla* the insertion is as described by Gibbs (1999) but we observed an additional origin from the plantar aponeurosis.

In *Pan* we observed two different patterns of insertion, both described by other authors: insertion onto digits 2, 3 and 4, and insertion onto digits 3 and 4 (Figure. 5) (Beddard, 1893, Champneys, 1872, Dwight, 1895, Hepburn, 1892, MacDowell, 1910; Miller, 1952; Sarmiento, 1983, Sutton, 1883). Innervation of the flexor digitorum brevis is by the medial plantar nerve (L_4-L_5) in modern humans, *Pan paniscus* and *Gorilla* (Raven, 1950; Miller, 1952; Gibbs, 1999).

Electromyographic analyses of modern humans have shown that this muscle is active when the strength of the long flexor of the digits is insufficient or when strength is needed at the interphalangeal joints of the toes in order to cling to a substrate (Reeser et al., 1983).

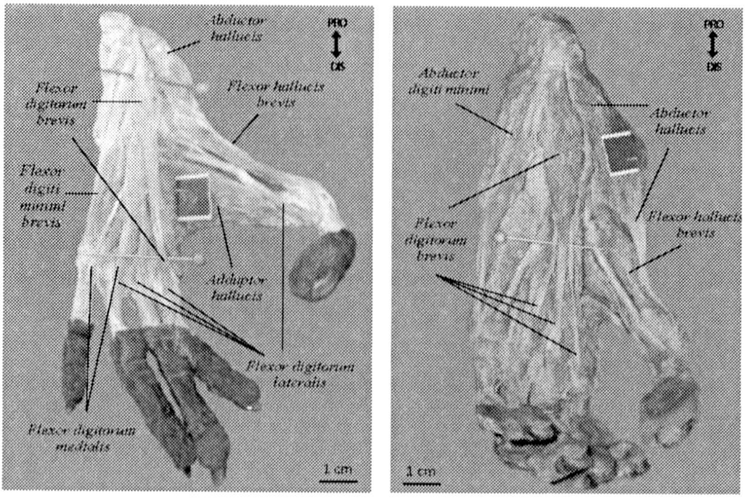

Figure 4. GWU HL1 (*Hylobates lar*, juvenile female; on the left) and HU HL1 (*H. lar*, adult male; on the right), plantar view of the left foot showing variation in the insertions of flexor digitorum brevis (in the specimen on the left onto digit 2; in the specimen on the right onto digits 2, 3 and 4).

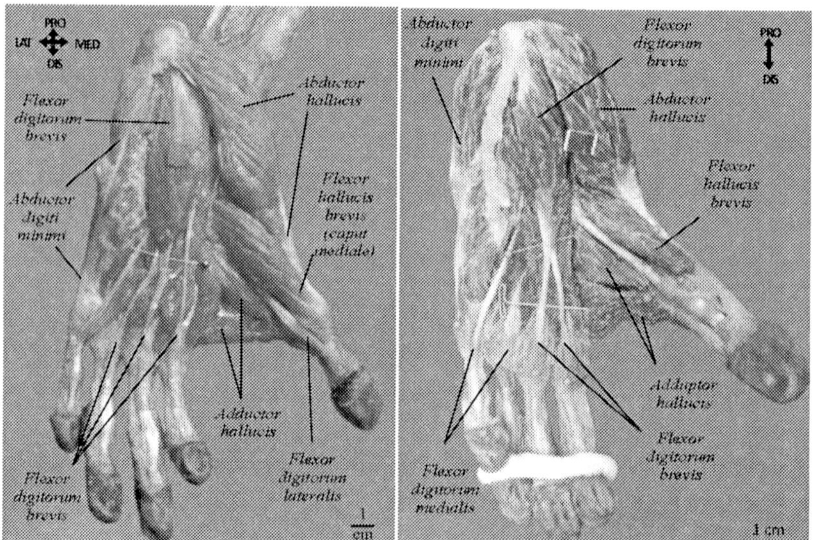

Figure 5. VU PT1 (*Pan troglodytes*, adult male; on the left) and GWU PT1 (*P. troglodytes*, adult female; on the right), plantar view of the right foot showing variation in the insertions of flexor digitorum brevis (in the specimens on the left onto digits 2, 3 and 4; in the specimen on the right onto digits 2 and 3).

In modern humans the *quadratus plantae* has two heads (i.e. one from from the lateral plantar surface of the calcaneus and the other from the medial plantar surface). In apes the muscle is often absent, and when present it is usually smaller than in modern humans and it has a single origin from the lateral aspect of the calcaneus. The double-headed muscle is described as unique for modern humans (Aiello and Dean, 1990), although Gibbs (1999) stated that the medial head of this muscle is usually present in *Pongo*, but not in *Gorilla* and *Pan*. The muscle is absent in the gibbons and orangutans dissected by us, but has been reported by previous authors in *Pongo* (Chapman, 1880; Hepburn, 1892; Hafferl, 1929; Lewis, 1962; Church, 1861-1862). In the gorilla and chimpanzees dissected by us the muscle inserts mainly onto the tendon of the flexor digitorum medialis that goes to digit 5. However exceptions for *Pan* are known, where the quadratus plantae merges with the tendons of the flexor digitorum medialis going to digits 3 and 4 (Dwight, 1895). Innervation of the quadratus plantae is by the lateral plantar nerve (L_4-L_5) in modern humans and in *Gorilla* (Raven, 1950; Warwick and Williams, 1973; Gibbs, 1999). It is commonly assumed that in all hominoids the function of the quadratus plantae is to straighten out the pull of the flexor digitorum medialis, adjusting its action along the longitudinal axes of the phalanges (Wood-Jones, 1949; Lewis, 1962). However, more recent studies have shown that (along with the flexor digitorum brevis) its main function is to supplement the action of the flexor digitorum medialis when powerful flexion is required (Reeser et al., 1983).

The origin of the *lumbricales* is from the flexor digitorum medialis tendons in all hominoids and, with the exception of modern humans, from the flexor digitorum lateralis tendons. The insertion in all taxa is onto the medial aspect of the base of the proximal phalanx of digits 2-5. Beddard's (1893) and Hepburn's (1892) descriptions of the first and second lumbricales in hylobatids were corroborated by our dissections. In the hylobatids dissected by us the lumbrical 3 merges with lumbrical 4, which inserts onto the medial side of the base of the proximal phalanx of digit 5. In *Pongo* the first lumbricales has a single head of origin, as described by other authors (Beddard, 1893; Boyer, 1935; Chapman, 1880; Hepburn, 1892; Sonntag, 1924), the second lumbricales had a dual origin in one of our orangutans, as reported by some authors (Beddard, 1893; Boyer, 1935; Chapman, 1880; Hepburn, 1892; Sonntag, 1924). Boyer (1935) reported a single specimen of *Pongo* in which the second lumbrical had a single origin, as found in one of our orangutans. The third and fourth lumbricales have a double origin as described in the literature (Beddard, 1893; Boyer, 1935; Hepburn, 1892; Sonntag, 1924).

Regarding *Gorilla*, the specimen dissected by us has the usual condition found in gorillas according to Gibbs (1999) (Table 4). Humphry (1866-1867) reported seven lumbricales in *Pan* while Beddard (1893) reported three and other authors reported four as found in our dissections (Beddard, 1893; Dwight, 1895; Hepburn, 1892; Miller, 1952; Sutton, 1883). Innervation of the lumbricales is by the medial plantar nerve (lumbrical 1) and lateral plantar nerve (lumbricales 2, 3 and 4) in modern humans (but this is prone to variation) and *Gorilla* (Raven 1950; Gibbs, 1999), but according to Miller (1952) in *Pan paniscus* lumbricales 1 and 2 are innervated by the medial plantar nerve and lumbricales 3 and 4 by the lateral plantar nerve. The lumbricales flex the metatarsophalangeal joints without flexing the interphalangeal joints (e.g., Aiello and Dean, 1990).

In modern humans the *adductor hallucis* is divided into two distinct portions. According to Keith (1929) in African apes the muscle exhibits an 'intermediate state' between modern humans (where its two heads are widely separated) and the powerful fan-shaped muscle that is present typically in modern human fetuses and in some non-anthropoids). However, the transverse head is always more developed in adult apes than in adult modern humans (Aiello and Dean, 1990). In all hominoids the origin of the muscle is from the plantar aspect of MII and MIII for all, and MIV for all except for *Pongo* and the modern human dissected by us; *Pan* also has an origin from the cuboid. In all of the hominoids we dissected we found an insertion of the muscle onto the base of the proximal phalanx of the hallux and, with the exception of modern humans, also onto MI head. Brooks (1887) suggested that in hylobatids the two heads of the muscle are always fused at the origin, but this was not the case in the gibbons dissected by us. Innervation of the adductor hallucis is by the lateral plantar nerve (S_1-S_2) in modern humans and in Asian apes, and in *Pan* it is by the deep branch of the lateral plantar nerve (Brooks, 1887; Hepburn, 1892; Sonntag, 1924; Raven, 1950; Miller, 1952; Gibbs, 1999). In *Pongo* and *Pan* an additional innervation from the medial plantar nerve has been reported (Brooks, 1887; Hepburn, 1892; Sonntag, 1924).

In all hominoids the *flexor digiti minimi brevis* originates from the base of MV. In all apes there is an additional origin from the plantar tarsometatarsal ligament; in gorillas there is also an origin from the base of MIV. The insertion in all hominoids is onto the proximal phalanx of digit 5, and onto the medial side of this digit in Asian apes. This muscle has not been described previously in hylobatids, but it is present in our hylobatids (Table 4). Our orangutans exhibited a condition similar to that described by previous authors (Boyer, 1935; Sonntag, 1924), plus they also had an origin from the plantar tarsometatarsal ligament. Rolleston (1867-1868) reported the absence of this muscle in *Pan*, contrary to other authors (Miller, 1952; Raven, 1950; Champneys, 1872; Straus, 1930); but the flexor digiti minimi brevis was found in the chimpanzees dissected by us. Some authors described the origin of this muscle in *Pan* as merging with that of the opponens digiti minimi and of the plantar interossei (Champneys, 1872); this was not the case in our chimpanzees. Innervation of the flexor digiti minimi brevis is by the lateral plantar nerve (S_1-S_2) in modern humans, *Gorilla* and *Pan paniscus* (Raven, 1950; Miller, 1952; Gibbs, 1999). The muscle mainly flexes digit 5.

Table 5. Total weight (in grams) of the specimens dissected and of the muscle mass of the hindlimb, and percentage value relative to the total body weight

	Hylobates	*Pongo*	*Gorilla*	*Pan*
Total weight (g)	6.700	72.000	98.000	81.000
Muscles lower limb - weight (g)	402,4	3111,92	3964,4	5757,15
Lower limb percentage (%)	6,006	4,322	4,045	7,108

START HERE The *flexor hallucis brevis* is a large muscle with two heads in modern humans. It is generally much smaller in apes, the exception being that Lewis (1964b) states that in *Pan* it is as broad as it is in modern humans, being an exception among the apes. However, our dissections contradict his statement. In modern humans flexor hallucis brevis has an single origin from the cuboid, in *Gorilla* it also originates from the calcaneus and from the first tarsometatarsal joint. Insertion in all hominoids is onto the lateral and medial aspects of the proximal phalanx of the hallux, and also onto the distal region of MI in apes, and additionally onto the metatarsophalangeal joint of the hallux in *Gorilla*. In *Hylobates* the muscle has two heads, the medial one having an origin from the cuneonavicular plantar ligament. Regarding *Pongo*, some authors (Beddard, 1893; Bischoff, 1870; Ruge, 1878b) described the medial head fusing with the adductor hallucis; Boyer (1935) reported a lateral head fused with the opponens hallucis, and Church (1861-1862) referred to an insertion of the medial head onto the plantar fascia (Church, 1861-1862); none of these variants were observed in our orangutans. Concerning *Pan*, Beddard (1893) reported a single origin of the medial head from the cuboid, some authors described an origin of the muscle from the lateral cuneiform and the metatarsophalangeal joint in this taxon (Brooks, 1887; Champneys, 1872; Duvernoy, 1853; Miller, 1952), and a fusion of the lateral head and the abductor hallucis was also reported (Brooks, 1887; Champneys, 1872; Miller, 1952); none of these variants were observed in our chimpanzees. Innervation of the muscle is by the medial plantar nerve (L_4-S_5) in all hominoids (Brooks, 1887; Champneys, 1871; Raven, 1950; Miller, 1952; Gibbs, 1999).

The name *opponens digiti minimi* is sometimes given to a muscle bundle that lies deep to the flexor digiti minimi brevis in *Pongo* and in modern humans, although the muscle is usually not recognized as an independent structure in modern human anatomical terminology (e.g., Terminologia anatomica, 1998). The muscle has been described in *Pan* and *Gorilla*, where it has an origin from the base of MV (Raven, 1950; Gibbs, 1999; Ferrero, 2011). When noted in modern humans, it is described as originating from the distal half of MV, and when described in *Pongo* it is blended with the fibularis longus tendon sheath and the plantar fascia. There is no common insertion for all hominoids. We did not find the opponens digiti minimi in our hylobatids, but we did locate it in our orangutans, although various authors state that the muscle is missing in *Pongo* (Boyer, 1935; Fick, 1895; Hepburn, 1892; Sonntag, 1924). We also found it in our gorillas and our chimpanzees. Champneys (1872) stated that in *Pan* the muscle originates together with the flexor digiti minimi brevis and with the plantar interosseous of digit 5, but in our chimpanzees it originated instead from the base of MV and from the interosseous ligaments. Innervation of the muscle is by the lateral plantar nerve in *Pan paniscus* and in *Gorilla* (Raven, 1950; Miller, 1952; Gibbs, 1999). Its action is to oppose digit 5.

Among hominoids, the *opponens hallucis* is said to be usually present only in *Pongo* (Bischoff, 1870; Boyer, 1935; Brooks, 1887; Champneys, 1872; Chapman, 1880; Church, 1861-1862; Huxley, 1864; Langer, 1879; Primorse, 1899; Vrolik, 1841; Warwick and Williams, 1973); Bischoff (1870) described it as a variant in hylobatids. However, in all of the specimens dissected by us it was absent. In *Pongo* individuals that have the muscle it is reported to be innervated by the medial plantar nerve (Brooks, 1887) and its function is to oppose the hallux.

It has been claimed that the main difference between the *dorsal interossei* of modern humans and of apes is that in the latter the reference axis of digit abduction is digit 3, not 2. Thus, in most primates digit 3 has a double insertion while in modern humans it is digit 2 that has a double insertion (Aiello and Dean, 1990).

However, according to our dissections hylobatids also have digit 2 as the reference axis for digit abduction.

Table 6. Weight (in grams) and percentage values of some hindlimbs muscles relative to the weight of the hindlimb musculature

Muscle	*Hylobates*		*Pongo*		*Gorilla*		*Pan*	
	Weight	%	Weight	%	Weight	%	Weight	%
Gluteus maximus	29,7	7,381	237,08	7,618	—	—	200.045	3,475
Gluteus medius	25,5	6,337	216,18	6,947	—	—	488,58	8,486
Gluteus minimus	2,3	0,572	19,225	0,618	—	—	44,42	0,772
Scansorius	3,3	0,820	29,565	0,950	—	—	—	—
Ischiofemoralis	17,1	4,250	155,32	4,991	—	—	308,59	5,360
Gemellus superior	0,1	0,025	2,785	0,089	—	—	3,215	0,056
Gemellus inferior	0,3	0,075	4,505	0,145	—	—	6,91	0,120
Obturator externus	5,5	1,367	22,005	0,707	—	—	63,06	1,095
Obturator internus	3	0,746	38,825	1,248	—	—	55,365	0,962
Piriformis	5	1,243	26,515	0,852	—	—	37,28	0,648
Quadratus femoris	3,9	0,969	33,59	1,079	—	—	12,58	0,219
Iliacus	16,5	4,1	103,51	3,326	—	—	276,9	4,81
Psoas major	16,2	4,026	136,82	4,397	—	—	195.31	3,392
Psoas minor			86,49	2,783	—	—	—	—
Articularis genus	—	—	—	—	—	—	—	—
Rectus femoris	12,1	3,007	94,645	3,041	143	3,607	136,755	2,375
Vastus intermedius	27,2	6,759	38,46	1,236	552	13.924	223,07	3,875
Vastus lateralis	36,2	8,996	132,65	4,263			514,88	8,943
Vastus medialis	18,2	4,523	66	2,121			265,12	4,605
Sartorius	11,6	2,883	38,76	1,246	98,5	2,485	119,175	2,070

Table 6. (Continued)

Muscle		*Hylobates*		*Pongo*		*Gorilla*		*Pan*	
		Weight	%	Weight	%	Weight	%	Weight	%
Tensor fasciae latae		—	—	22,305	0,717	—	—	49,07	0,852
Adductor brevis		2,1	0,522	39,78	1,278	1125,5	28,390	113,225	1,967
Adductor longus		3,6	0,895	61,29	1,970			389,09	6,758
Adductor magnus		54,4	13,519	385,015	12,372			498,2	8,654
Adductor minimus		—	—	—	—			—	—
Gracilis		10,9	2,709	167,695	5,389	279	7,038	250,745	4,355
Pectineus		3,4	0,845	17,705	0,569	48	1,211	65,315	1,135
Biceps femoris	*Caput longum*	11,9	2,709	93,565	3,007	—	—	160,455	2,787
	Caput brevis			46,09	1,481	—	—	131,085	2,277
Semimembranosus		9,9	2,460	114,745	3,687	206	5,196	163,755	2,844
Semitendinosus		13	3,231	184,98	5,944	333,2	8,405	200,435	3,481
Extensor digitorum longus		4,5	1,118	48,24	1,550	61,6	1,554	75,245	1,307
Extensor hallucis longus		2,4	0,596	8,33	0,268	25,8	0,651	25,485	0,443
Tibialis anterior		9,2	2,286	69,525	2,234	127	3,204	121,425	2,109
Fibularis brevis		2,7	0,671	24,765	0,796	40,3	1,017	45,98	0,799
Fibularis longus		4,2	1,044	39,695	1,276	69,6	1,756	96.05	1,668
Gastrocnemius	*Caput laterale*	15,2	3,777	53,24	1,711	190,9	4,815	194,22	3,374
	Caput mediale			73,565	2,364			121,05	2,116
Plantaris		2,5	0,621	—	—	—	—	11,355	0,197
Soleus		8,8	2,187	98,93	3,179	159,5	4,023	218,325	3,792

Muscle		*Hylobates*		*Pongo*		*Gorilla*		*Pan*	
		Weight	%	Weight	%		Weight	%	Weight
Flexor digitorum medialis		6,5	1,615	73,825	2,372	86,4	2,179	67,2	1,167
Flexor digitorum lateralis		12,9	3,206	74,94	2,408	177	4,465	139,855	2,429
Popliteus		2,4	0,596	27,55	0,885	—	—	47,025	0,817
Tibialis posterior		2,4	2,087	30,38	0,976	58,2	1,468	131,155	2,278
Extensor digitorum brevis		1,7	0,422	27,355	0,879	14,2	0,358	23,43	0,407
Extensor hallucis brevis		0,6	0,149	3,615	0,116	5,1	0,129	8,155	0,142
Abductor digiti minimi		—	—	14,635	0,470	—	—	11,625	0,202
Abductor hallucis		2,6	0,646	4,59	0,147	34,4	0,868	40,555	0,704
Abductor metatarsi quinti		—	—	2,83	0,091	—	—	—	—
Flexor digitorum brevis		0,7	0,174	30,145	0,969	28	0,706	20,375	0,354
Quadratus plantae		—	—	—	—	—	—	2,83	0,049
Lumbricalis		1,4	0,348	9,875	0,317	7,9	0,199	7,78	0,135
Adductor hallucis	*Caput obliquum*	3,1	0,770	12,915	0,415	20,9	0,527	12.23	0.212
	Caput transversum			10,62	0,341	23,3	0,588	27,975	0,486
Flexor digiti minimi brevis		0,9	0,224	5,735	0,184	—	—	4,29	0,075
Flexor hallucis brevis	*Caput mediale*	—	—	7,48	0,240	15,9	0,401	13,83	0,240
	Caput laterale			8.84	0.284				
Opponens digiti minimi		—	—	0,745	0,024	—	—	1,01	0,018
Opponens hallucis		—	—	—	—	—	—	—	—
Interossei dorsales		1,5	0,373	37,495	1,205	20,1	0,507	18,79	0,326
Interossei plantares		—	—	17,33	0,557	13,1	0,330	16,52	0,287

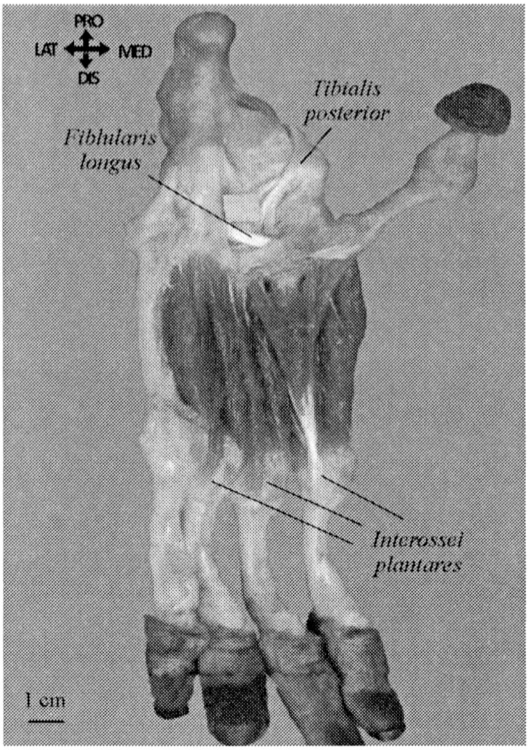

Figure 6. VU PP1 (*Pongo pygmaeus*, adult female), plantar view of the deepest muscles of the right foot, showing that the three interossei plantares.

The origin of the dorsal interossei in all hominoids is from the lateral surface of MII, MIII and MVI (from MI in all hominoids except *Gorilla*), and from the medial side of MII, MIII, MIV and MV; an additional origin from the medial cuneiform is found in *Pongo* and *Gorilla*. In *Pongo* and *Hylobates* the head of origin from MI is rudimentary or absent (Brooks, 1887); it was present in only one of our gibbons. Innervation of the dorsal interossei is by the lateral plantar nerve (S_1-S_2) in modern humans, and by the deep branch of the lateral plantar nerve in *Pan paniscus* (Raven, 1950; Miller, 1952; Gibbs, 1999). In all hominoids the dorsal interossei act as abductors of the digits and as flexors at the metatarsophalangeal joints.

In apes the *interossei plantares* originate from the medial surface of MII, and in *Pongo* also from the MI and from the navicular. Some authors (Boyer, 1935; Duvernoy, 1855-1856; Sonntag, 1924) report a single head of origin for each plantar interosseous in *Pongo*. One of our orangutans has a third plantar interosseous with a single head of origin from the medial side of the base of

MV. The second plantar interosseous originates from the lateral side of the base of MIII and from the medial side of MIV. The first plantar interosseous of the left hindlimb of the same individual has four heads of origin: from the lateral surface of the base of MI, the distal and medial regions of the navicular, the base of MII, and the medial surface of the base of MIII. On the right hindlimb all these origins are present except for that from MI (Figure. 6). Only modern humans have an insertion onto the proximal phalanx of digit 3, and only hylobatids have an insertion onto the medial side of the proximal phalanx of digit 2 that is not found in other hominoids. Innervation of the interossei plantares is by the lateral plantar nerve (S_1-S_2) in modern humans and by the deep branch of the lateral plantar nerve in *Pan paniscus* and *Gorilla* (Raven, 1950; Miller, 1952; Gibbs, 1999). The action of the interossei plantares is to adduct the digits and to flex them at the metatarsophalangeal joints.

Overview

Within the hominoid specimens dissected by us the scansorius, articularis genus, adductor minimus, fibularis tertius, quadratus plantae, opponens digiti minimi and opponens hallucis are usually missing in hylobatids, which usually have 60 lower limb muscles (including, as is the case in other hominoids, 4 interossei dorsales, 4 lumbricales and 3 interossei plantares).

The articularis genus, adductor minimus, fibularis tertius, quadratus plantae and opponens hallucis are usually missing in orangutans, which thus usually have a total of 62 lower limb muscles.

The scansorius, gemellus superior, fibularis tertius, plantaris, and opponens hallucis are usually absent in gorillas, which usually have 62 lower limb muscles.

The scansorius, articularis genus, adductor minimus, fibularis tertius, abductor metatarsi quinti, and opponens hallucis are usually missing in chimpanzees, which usually have 61 muscles.

In modern humans the scansorius, ischiofemoralis, abductor metatarsi quinti, adductor minimus, opponens digiti minimi and opponens hallucis are usually missing, so there are usually 61 muscles.

Thus modern human upright bipedalism is not related to an increase in the number of lower limb muscles or to the presence of any autapomorphic muscles, but instead to differences in the general configuration, attachments and/or mass of these structures.

References

Aiello L and Dean C . 1990. An introduction to human evolutionary anatomy. Academic Press. London.

Alexander R McM. 1984. Walking and running. *American Scientist*, 72, p348-354.

Ankel-Simons F. Primate anatomy. An introduction. Academic Press. London.

Appleton AB and Ghey PHR. 1929. An example of the cervico-costo-humeral muscle of Gruber. *J. Anat.*, 63: 434-436.

Barnard WS. 1875. Observations on the membral musculation of *Simia satyrus* (Orang) and the comparative mycology of man and the apes. *Proc. Am. Assoc. Sci.*, 24, p112-144.

Beddard FE 1893. Contributions to the anatomy of the anthropoid apes. *TansZool. Soc. Lond.*, 13, p177-218.

Bischoff TLW. 1870. Beiträge zur Anatomie des *Hylobates leuciscus* and zu einer vergleichenden Anatomie der Muskeln der Affen und des Menschen. Abh.Bayer.Akad.Wiss. München. *Math. Phys. Kl.*, 10, p197-297.

Blake ML. 1976. The queantitative myology of the hind limb of Primates with special reference to their locomotor adaptations. PhD Thesis, Magdalene College, University of Cambridge.

Boyd R and Silk JB. 2001. Cómo evolucionaron los humanos. Ariel. Barcelona.

Boyer EL. 1935. The musculature of the inferior extremity of the orang-utan Simia satyrus. *Am. J. Anat.*, 56(2), p192-256.

Brooks HSJ. 1887. On the short muscles of the pollex and hallux of the anthropoid apes, with special reference to the opponens hallucis. *J. Anat. Physiol.*, 22:78-95.

Brown B. 1983. An evaluation of primate caudal musculature in the identification of the ischiofemoralis muscle. *Am. J. Phys. Antropol.*, 60, p177-178 (Abst).

Camper P. 1779. Account of the organs of speech of the Orang Outang. *Phil. Trans. Roy. Soc.*, 69, p139-159.

Camper P. 1782. Naturkundige verhandelbingen over den Orang-Outang; en eenige andere Aapsoorten. Over den Ehinoceros met den dubbelen horen; en over het Rendierr. Amsterdam: Meijer and Wamars.

Champneys F. 1872. The muscles and nerves of a Chimpanzee (Troglodytes niger) and a Cynocephalus anubis. *J. Anat. Physiol.*, 6:176-211.

Chapman HC. 1879. On the structure of the chimpanzee. *Proc. Zool. Soc. Lond.*, p52-63.

Chapman HC. 1880. On the structure of the orang outing. *Proc. Zool. Soc. Lond.*, p160-174.

Chapman HC. 1887. On the structure of the gorilla. *Proc. Zool. Soc. Lond.*, p385-394.

Chiarelli B. 1963. Comparative morphometrics of primate chromosomes. III. The chromosomes of the genera Hylobates, Colobus and Presbytis. *Caryologia*, 16, p637-648.

Church WS. 1861-1862. On the myology of the orang utang (Simian morio). *Nat. Hist. Review.*, 1, p510-516; 2, p89-94, 414-423.

Coppens Y. 1983. Le singe, L´Afrique et l´homme. Ed. Fayard. Paris.

Crovella S, Montagno D and Rumpler Y. 1992. Highly repeated DNA analysis and systematic of the Lemuridae, a family of Malagasy prosimians. *Primates*, 34, p61-69.

Day M. 1986. Bipedalism: Pressures, Origins and Modes. In: Wood B, Martin L, Andrews P (eds.) 1986. Major topics in primate and human evolution. Cambridge University Press (New York).

Deniker J. 1885-1886. Recherches anatomiques et embryologiques sur les singes anthropoïdes: Foetus de gorille et de gibbon. *Arch. Zool. Exp. Gen. Ser.*, 2, 3(3), pl-265.

Duckworth WLH. 1898. Note on a foetus of Gorilla savage. *J. Anat.*, 33, p82-90.

Duvernoy M. 1853. Cited in Brooks (1887).

Duvernoy M. 1855-1856. Des caractéres anatomiques de grands singes pseudoanthropomorphes. *Arch. Mus. Natl. Hist. Nat.,* (Paris). 8, p1-248.

Dwight T. 1895. Notes on the dissection and brain of the chimpanzee "Gumbo". Mem. Boston *Soc. Nat. Hist.*, 5, p31-51.

Eisler P. 1895. Die Modern humanslogie der Extremitaeten. Abh.Naturforsch.*Ges. Halle.*, 19, p87-346.

Elftman H and Atkinson WB. 1950. The abdominal viscera of the gorilla. In: The Anatomy of the Gorilla (ed. Gregory WK), p197-201. New York: Columbia University Press.

Embleton. 1864. Cited in Chapman (1879).

Fabre P-H, Rodrigues A, Douzery EJP. 2009. Patterns of macroevolution among Primates inferred from a supermatrix of mitochondrial and nuclear DNA. *Mol. Phyl. Evol.*, 53, p808–825.

Feneis H. 2000. Nomenclatura anatómica ilustada. Ed. Masson. Barcelona.

Ferrero EM. 2011. Anatomía comparada del sistema muscular de la extremidad posterior en primates superiores. PhD Thesis, The University of Valladolid, Spain.

Fick R. 1895. Vergleichend-anatomische Studien an einem erwachsenen Orang-utang. *Arch. Anat. Physiol. Anat. Abt.*, p1-100.

Fleagle JG. 1992. Primate locomotion and posture. In The Cambridge Encyclopedia of Human Evolution (ed S. Jones, R. Martin, D. Pilbeam), pp. 75-78. Cambridge Univertisy Press, Cambridge.

Forster A. 1903. Die Insertion des Musculus semimembranosus. Arch.Anat.Physiol. *Anat. Abt.,* p257-320.

Frey H. 1913. Der Musculus triceps surae in der Primatenreihe. *Morphol. Jaharb.*, 47, p1-192.

Fürst CM. 1903. Der Musculus Popliteus und seine Sehne. E. Malstrom, Lund.

Gebo DL. 1989. Locomotor and phylogenetic considerations in anthropiod evolution. *J. Human Evol.,* 18, p201-233.

Gibbs S, Collard M and Wood B. 2002. Soft-tissue anatomy of the extant hominoids: a review and phylogenetic analysis. *J. Anat.*, 200, p3-49.

Gibbs S. 1999. Comparative soft tissue morphology of the extant *Hominoidea*, including Man. Unpublished PhD Thesis, The University of Liverpool, Liverpool.

Goodman M, Bailey WJ, Hayasaka K, Stanhope MJ, Slighton J and Czelusniak J. 1994. Molecular evidence on primate phylogeny from DNA sequences. *Am. J. Phys. Anthropol.,* 94, p3-24.

Goodman M. 1963. Man's place in the phylogeny of the primates as reflected in serum proteins. In: Classification and Human Evolution (Washburn SL, ed), p204-234. Chicago: Aldine.

Gray EG and Basmajain JV. 1968. Electromyography and cinematography of leg and foot ("normal" and fat) during walking. *Anatomical Record.,* 161, p1-5.

Gray JE. 1825. Outline of an attempt at the disposition of the Mammalia into tribes and families with a list of the genera apparently appertaining to each tribe. *Ann. Phil. N.S.* 10, p337-344.

Groves C. 2001. Primate taxonomy. Smithsonian Institution Press. Wasington and London.

Hafferl A. 1929. Bau und Funktion des Affenfusses. Ein Beitrag sur Gelenk und Muskel-mechanik. *Z.Anat. Entwicklungsgesch.*, 88, p749-783.

Hamada Y. 1985. Primate hip and thigh muscles: comparative anatomy and dry weights; In:Kondo, S(ed); Primate Morphophysiology. Locomotor Analyses and Human Bipedalism. University of Tokyo Press, Tokyo. p131-152.

Hartmann R. 1883. Cited in MacDowell (1910).

Hecker P. 1922. Formation du peronier anterieur chez un chimpanzee. *Arch. Anat. Histol. Embryol.*, 1, p147-155.

Hepburn D. 1892. The comparative anatomy of the muscles and nerves of the superior and inferior extremities of the anthropoid apes. *J. Anat. Phys.*, 26, p149-186, 324-356.

Humphry GM. 1866-1867. On some points in the anatomy of the chimpanzee. *J. Anat. Phys.,* 1, p254-268.

Huxley TH. 1864. Lectures on the structure and classification of the Mammalia. *Med. Times and Gaz.*, (London). 1, p564-565.

Jungers WL, Meldrum DJ and Stern JT. 1993. The functional and evolutionary significance of the human peroneus tertius muscle. *J. Hum. Evol.*, 25, p377-386.

Kaneff A. 1986. Die Aufrichtung des Menschen und die morphologisches Evolution der Musculi extensors difitorum pedis unter dem Gesichtpunkt der evolutiven Myologie. Teil I. *Gegenbaur's Morphol. Jahrb.*, 132, p375-419.

Kaplan EB. 1958a. The iliotibial tract. Clinical and morphological significance. *J. Bone. Jt. Surg.*, 40A, p817-832.

Kaplan EB. 1958b. Comparative anatomy of the extensor digitorum longus in relation to the knee joint. *Anat. Rec.*, 131 (2), p129-149.

Kaseda M, Nakamura M, Ischihara N, Hayakawa T and Asari M. 2008. A macroscopic examination of M. Biceps Femoris and M. Gluteus Maximus in the Orangutan. *J. Vet. Med. Sci.*, 70(3), p217-222.

Keith A. 1894. The myology of the catarrhini: A study in evolution. PhD Thesis. University of Aberdeen.

Keith A. 1926. The Engines of the Human Body. Philadelphia: J. B. Lippincott.

Keith A. 1929. The history of the human foot and its bearing on orthopaedic practice. *J. Bone Joint Surg.*, 11, p10-32.

Kimura K and Takahashi Y. 1985. The peroneus tertius muscle in the crab-eating monkey (Macaca fascicularis). *Okajimas Folia Anat. Jpn.*, 62, p173-186.

Kohlbrügge JHF. 1890-1891. Versuch einer Anatomie des Genus *Hylobates*. In: Weber M (ed.), Zoologische Ergebnisse Einer Reise in Niederländisch Ost-Indien. Leiden: Verlag von EJ Brill; pp. 211-354 (Vol. 1), 138-208 (Vol. 2).

Kumakura H. 1989. Functional analysis of the biceps femoris muscle during locomotor behavior in some primates. *Am. J. Phys. Anthropol.*, 79, p379-391.

Langer C. 1879. Die Muskeln der Extremitäten des Oran gals Grundlage einer vergleichend-anatomischen Untersuchung. Sitzungsber.Kaiserl.Akad. Wiss.Wien. Math-Naturwiss.Cl, 80, p177-222.

Latimer B and Lovejoy CO. 1989. The calcaneus of *Australopithecus afarensis* and its implications for the evolution of bipediality. *Am. J. Phys. Anthropol.*, 78, p369-386.

Latimer B, Ohman JC, and Lovejoy CO. 1987. Talocrural joint in African hominoids: Implications for *Australopithecus afarensis. Am. J. Phys. Anthropol.*, 74, p155-175.

Le Comte LD. 1697. Novesux Mémores sur L'etat Présent de la China, II, p361-364. Amsterdam.

Le Double F. 1897. Traité des variations du système musculaire de l'homme et leur signification au point de vue de l'anthropologie zoologique, I et II. Paris. Librairie C. Reinwald.

le Minor JM. 1992. Brief communication: The popliteal sesamoid bone (cyamella) in primates. *Am. J. Phys. Anthropol.*, 87, p107-110.

Lewis OJ. 1962. The morphology of M. flexor accessories and the associated long flexor tendons. *J.Anat.*, 96, p321-333.

Lewis OJ. 1964a. The evolution of the long flexor muscles of the leg and foot. *Int. Rev. Gen. Zool.,* 1, p165-185.

Lewis OJ. 1964b. The tibialis posterior tendon in the primate foot. *J. Anat.,* 98, p209-218.

Lewis OJ. 1966. The phylogeny of the cruropedal extensor musculature with special reference to the primates. *J. Anat.*, 100, p865-880.

Lewis OJ. 1980. The joints of the evolving foot. Part II. The intrinsic joints. *Journal of Anatomy*, 130, p833-857.

Lewis OJ. 1989. Functional Morphology of the Evolving Hand and Foot. Clarendon Press, Oxford.

Loth E. 1908. Die Aponeurosis plantaris in der Primaten reihe. Gegenbaurs *Morphol. Jahrb.*, 38, p194-322.

Loth E. 1914. Zur Anthropologie der Plantaraponeurose. *Morphol. Jahrb.*, 48, p83-96.

Loth E. 1931. Anthropologie des parties molles (muscles, intestins, vaisseaux, nerfs peripheriques). Mianowski-Masson et Cie, Varsovie, Paris.

Lovejoy CO. 1981. The origin of man. *Science*, 211, p341-350.

MacAlister A. 1871. On some points in the myology of the chimpanzee and others of the primates. *Ann. Mag. Nat. Hist.*, 7:341-351.

MacAlister A. 1873. The muscular anatomy of the gorilla. *Proc. Royal Irish Acad.*, 11, p501-506.

MacDowell EC. 1910. Notes on the myology of *Anthro-popithecus niger* and *Papio-thoth ibeanus*. *Am. J. Anat.*, 10, p. 431-460.

MacMinn RMH 1990. Last's Anatomy: regional and applied; 8th ed. Churchill Livingstone, Edinburgh.

Mann R and Inman VT. 1964. Phasic activity of intrinsic muscles of the foot. *Journal on Bone and Joint surgery*, 46A, p469-481.

Manners-Smith T. 1908. A study of the cuboid and os perineum in the primate foot. *J. Anat. Phys.*, 42, p397-414.

Martin RD. 1990. Primate origins and evolution: A phylogenetic analysis. Princeton University Press. Princeton. N.J.

Marzke MW, Longhill JM and Rasmussen SA. 1988. Gluteus maximus muscle function and the origin of hominid bipedality. *Am. Phys. Anthropol.*, 77, p519-528.

Masters JC, Rayner RJ, Ludewick H, Zimmerman E, Molez-Verrière N, Vincent F and Nash T. 1994. Phylogenetic relationships among the *Galaginae* as indicated by erythrocytic allozymes. *Primates*,35, p177-190.

Merker B. 1984. A note on hunting and hominid origins. *Amer. Anthropol.*, 86, p112-114.

Michaëlis P. 1903. Beiträge zur vergleichenden Myologie des *Cynocephalus babuin*, Simia satyrus, Trohlodytes niger. *Arch. Anat. Phys. Anat.Abt.*, p205-256.

Miller RA. 1945. The ischial callosities of primates. *Am. J. Anat.*, 76, p67-87.

Miller RA. 1952. The musculature of Pan paniscus. *Am. J. Anat.*, 91(2), p182-232.

Moore KL and Dalley AF. 2002. Anatomía con orientación clínica. Editorial Medica Panamericana.

Morton DJ. 1922. Evolution of the human foot. Part 1. *Am. J. Phys. Anthropol.*, 5, p305-336.

Mysberg WA. 1917. Über die Verbinderungen zwischen dem Sitzbeine und der Wirbelsäule bei den Säugetieren. *Anat. Hefte*. 54, p641-668.

Nuttall GHF. 1904. Blood Immunity and Blood Relationship. Cambridge: Cambridge Universit Press.

Olson TR. 1997. A.D.A.M. Atlas de anatomía humana. Masson – Williams and Wilkins.

Owen R. 1830-1831. On the anatomy of the orangutan (Simia satyrus, Linn.). *Proc. Zool. Soc. Lond.*, 1:4-5, 9-10, 28-29, 66-72.

Owen R. 1843. Notes of the dissection of a female orang-utan (Simia satyrus, Linn.). *Proc. Zool. Soc.*, London. 11, p123-124.

Owen R. 1846. Notes on the dissection of a female chimpanzee. *Proc.Zool.Soc.*, London. 14, p2-3.

Owen R. 1849. Osteological contributions to the natural history of the chimpanzees (Troglodytes, Geoffroy) including the description of the skull of a large species (Troglodytes gorilla, Savage) discovered by Thomas S. Savage, MD in the Gabon country, West Africa. *Trans.Zool.Soc.*,London. 3, p381-422.

Owen R. 1859. On the gorilla (Troglodytes gorilla, Sav.). *Proceedings of the Sool. Soc.*, Lond. p1-23.

Owen R. 1865. Memoir on the gorilla (Troglodytes gorilla, Savage). London: Tylor, Francis.

Owen R. 1868. The anatomy of vertebrates. Vol. 3, Mammals. London: Longmans, Green and Co.

Oxnard CE and Lisowski FP. 1980. Functional articulation of some hominoid foot bones: Implications for the Olduvai (Hominid 8) foot. *Am. J. Phys. Anthropol.*, 52, p107-117.

Pearson K and Davin AG. 1921. On the sesamoids of the knee joint. Biometrika. 13, p133-175, 350-400.

Potau JM, Artells R, Bello G, Muñoz C, Monzó M, Pastor JF, de Paz F, Barbosa M, Diogo R and Wood B. 2001. Expression of Myosin Heavy Chain Isoforms in the Supraspinatus Muscle of Different Primate Sepecies: Implications for the Study of the Adaptation of Primate Shoulder Muscles to Different Locomotor Modes. *Int. J. Primatol.*, p1-14.

Prejzner-Morawska A y Urbanowicz M. 1971. The biceps femoris muscle in lemurs and monkeys. *Folia Morphol.*, 30, 9465-482.

Primrose A. 1899. The anatomy of the Orang-outang (*Simia satyrus*), an account of some of its external characteristics, and the myology of the extremities. *Trans Royal Can. Inst.,* 6:507-594.

Primrose A. 1900. The anatomy of the orang-outang. Univ. of Toronto Studies. *Anat.series*, 1, p1-94.

Ranke K. 1897. Muskel-und Nervenvariationen der dorsalen elemente des Plexus ischiadicus der Primaten. *Arch. Anthropol.,* 24, p117-144.

Raven HC. 1950. Regional anatomy of the *Gorilla*. In: Gregory WK (ed.), The anatomy of the *Gorilla*. Columbia University Press, New York, p. 15-188.

Redmond I. 2008. Primates of the world. The amazing diversity of our closest relatives. New Holland Publishers. London.

Reeser LA, Susman RL and Stern JT. 1983. Electromyographic studies of the human foot: experimental approaches to hominid evolution. *Foot and Ankle*, 3, p391-407.

Robinson JT, Freedman L and Sigmon BA. 1972. Some aspects of pongid and hominid bipedality. *J. Hum. Evol.*, 1, p361-369.

Rodman PS y McHenry HM. 1980. Bioenergetics and the origin of hominoid bipedalism. *Am. J. Phys, Anthropol.*, 52, p103-106.

Rolleston G. 1867-1868. Cited in Champneys (1872).

Rowe N. 1996. The pictorial guide to the living primates. Pogonias Press. East Hampton. New York.

Rózycki S. 1922. Cited in Urbanowicz and Prejzner-Morawska (1972).

Ruge G. 1878a. Untersuchung uber die Extensorengruppe aus Unterschenkel und Füsse der Säugethiere. *Morphol. Jahrb.*, 4, p592-643.

Ruge G. 1878b. Zur vergleichenden Anatomie der tiefen Muskeln in der Fusssohle. *Morphol. Jaharb.*, 4, p644-659.

Ruvolo M. 1997. Molecular phylogeny of the hominoids: inference from multiple independent DNA sequence data sets. *Mol. Biol. Evol.,* 14, p248-265.

Sandifort G. 1840. Ontleedkundige beschouwing van een' volwassen Orang-oetan (Simia satyrus, Linn.) van het mannelijk geslacht. In: Temminchk, CJ; Verhandelingen over de Natuurlijke geschiedenis der Nederlandische overzeesche bezittingen, door de leden der Natuurkundige commissie in Indië en andere Schrijvers (Zoologie). 3, p29-56, Leiden.

Sarich VM. 1968. Immunological time scale for hominid evolution. *Science*, 158 p1200.

Sarmiento EE. 1983. The significance of the heel process in anthropoids. *Int. J. Primatol.*, 4, p127-152.

Savage TS and Wyman J. 1847. Notice of the external characters and habits of Troglodytes gorilla, a new species of orang from the Gaboon River. Mem. Boston Soc. Natural History (*A.K.A. Boston Journal of Natural History*). 5, p417-443.

Schultz AH. 1979. Los primates. Ed. Destino. Barcelona.

Sigmon BA and Farslow DL. 1986. The primate hindlimb. In : Comparative Pimate Biology, Vol. 1: Systematics, Evolution and Anatomy (DR Swindler and J Erwin, Eds), p671-718. New York: A. Liss.

Sigmon BA. 1969. The scansorius muscle in pongids. *Primates*, 10, p247-261.

Sigmon BA. 1974. A functional analysis of pongid hip and thigh musculature. *J. Hum. Evol.*, 3, p161-185.

Sneath RS. 1955. The insertion of the biceps femoris. *J. Anat.*, 89, p550-553.

Sonntag CF. 1924. On the anatomy, physiology, and pathology of the orang-outan. *Proc. Zool. Soc. Lond.*, 24, p349-450.

Stern JT and Susman RL. 1983. The locomotor anatomy of *Australopithecus afarensis*. *Am. J. Phys. Anthropol.*, 60, p279-317.

Stern JT. 1972. Anatomical and functional specializations of the human gluteus maximus. *Anta. J. Phys. Anthropol.*, 36, p315-340.

Straus WL Jr. 1930. The foot musculature of the highland gorilla (*Gorilla beringei*). *Q. Rev. Biol.*, 5, p261-317.

Straus WL Jr. 1930. The foot musculature of the highland gorilla (*Gorilla beringei*). *Q. Rev. Biol.*, 5, p261-317.

Sutton JB. 1883. On some points in the anatomy of the chimpanzee (*Anthropopithecus troglodytes*). *J. Anat. Physiol.*, 18:66-85.

Swindler DR and Wood CD. 1973. An atlas of primate gross anatomy; baboon, chimpanzee and man. University of Washington Press, Seattle.

Tappen NC. 1955. Relative weights of some functionally important muscles of the thigh, hip and leg in a gibbon and in man. *Am. J. Phys. Anthropol.*, 13(3), p415-420.

Terminologia Anatomica, 1998. Federative commitee on Anatomical Terminology. Stuttgard: Georg Thieme.

Testut, L. 1884. Les anomalies musculaires chez l´homme expliquèes par l´anatomie comparée et leur importance en anthropologie. Paris: Masson.

Traill T. 1817. Anatomy of a chimpanzee. Mem.Wernerian *Nat. Hist. Soc. Edinb.*, 3, p1-149.

Tulp N. 1641. Observationum Medicarum, *Libre Tres*, p274-279.

Tyson E. 1699. Orang-utan, sive Modern humans Sylvestris: or, The anatomy of a pygmie compared with that of a monkey, an ape and a man. To which is added, a philological essay concerning the pygmies, the cynocephali, the satyrs, and sphinges of the ancients. Wherein it Will Appear That They Are All Either Apes or Monkeys, and Not Men, as Formerly Pretended. London: Bennet.

Urbanowicz M and Prejzner-Morawska A. 1972. The triceps surae muscle in chimpanzees. *Folia Morphol.*, 31, p433-440.

Vallois H. 1914. Étude anatomique de l'articulation du genou ches les Primates. L'Abeille,Montpellier.

van den Broek AJ. 1914. Studien zur Morphologie des Primatenbeckens. *Morphol. Jahrb.*, 49, p1-118.

van Westrienen A. 1907. Das Kniegelenk der Primaten, mit besonderer Berücksichtigung der Anthropoiden. Petrus Camper. 4, p1-60.

Vereecke E and Aerts P. 2008. The mechanics of the gibbon foot and its potential for elastic energy storage during bipedalism. *J. Exp. Biol.*, 211, p3661-3670.

Vereecke E, D'Août K, Payne R and Aets P. 2005a. Functional analysis of the foot and ankle myology of gibbons and bonobos. *J. Anat.*, 206, p453-476.

Vereecke E, D'Août K, Van Elsacker L, De Clercq D and Aerts P. 2005b. Functional analysis of the gibbon foot during terrestrial bipedal walking: plantar pressure distributions and three-dimensional ground reaction forces. *Am. J. Phys. Anthropol.*, 128, p659-669.

Vrolik W. 1841. Recherches d'anatomie comparée, sur le chimpancé. Johannes Miller, Amsterdam.

Warburn SL. 1950. Thoracic viscera of the gorilla. In: The Anatomy of the *Gorilla* (ed. Gregory WK), p189-195. New York: Columbia University Press.

Warwick P and Williams PL. 1973. Gray's Anatomy. 35th British Edition. Saunders Co., Philadelphia.

Waterman HC. 1929. Studies on the evolution of the pelvis of man and other primates. *Bull. Am. Mus. Nat. Hist.*, 55, p585-642.

White TD and Suwa G. 1987. Hominid footprints at Laetoli: Facts and interpretations. *Am. J. Phys. Anthropol.*, 72, p485-514.

Wilder B. 1862. Contributions to the comparative myology of the chimpanzee. Boston *J. Nat. Hist.*, 6, 352-384.

Wilson DE and Reeder DM. 2005. Mammal species of the world. A taxonomic and geographic reference. Vol. 1. 3nd ed. The Johns Hopkins University Press. Baltimore.

Wood-Jones F. 1949. Structure and Function as Seen in the Foot, 2nd edn. London: Bailliére-Tindall.

Xu X and Arnason U. 1996. The mitocondrial DNA molecule of Sumatran orangutan and molecular proposal for two (Bornean and Sumatran) species of orangutan. *J. Molec. Biol.*, 43, p431-437.

Yirga S. 1987. Interrelation between ischium, thigh extending muscles and locomotion in some primates. *Primates*, 28(1), p79-86.

Zihlman A and Brunker L. 1979. Hominid bipedalism: Then and now. *Yearb. Phys. Anthropol*, 22, p132-162.

In: Primates
Editors: E. F. Hughes and M. E. Hill
ISBN: 978-1-62081-498-7

Chapter II

Alzheimer Pathology in Non-Human Primates and its Pathophysiological Implications

A. Toledano[1,*], M. I. Álvarez[1], P. Carmona[2], A. Toledano-Díaz[3] and C. I. Fernández-Verdecia[4]

[1] Instituto Cajal, CSIC, Madrid, Spain
[2] Instituto de Estructura de la Materia, CSIC, Madrid, Spain
[3] Departamento de Reproducción Animal, INIA, Madrid, Spain
[4] Centro Internacional de Restauración Neurológica (CIREN), Cubanacán, La Habana, Cuba

Abstract

Alzheimer's disease (AD) is commonly believed exclusive to humans, even though amyloid accumulation has been detected in senile individuals of all the sub-orders, families, genera and species of non-human primates studied to date. Indeed, in some species, all senile individuals show such deposits. However, only rarely are they associated with neuropathological abnormalities. Further, the highly phosphorylated tau protein seen in humans with AD is only very rarely encountered in non-human primates. In fact it is only detected with any regularity in

* Correspondence: Dr. Adolfo Toledano, Instituto Cajal, C.S.I.C. Avd Dr. Arce 37, 28002 Madrid, Spain. E-mail: atoledano@cajal.csic.es.

aged mouse lemurs (*Microcebus murinus*), and the changes related to its accumulation appear to be of very little significance. This chapter discusses the pathophysiological meaning of the results obtained in the neuropathological study of the brains of higher (*Chlorocebus aethiops*, *Macaca mulata* and *Macaca arctoides* among others) and lower (Microcebus murinus among others) non-human primates. In macaques, the existence of a possible continuum between the "normal" ageing process, "normal" ageing with no profound neuropathological or cognitive-behavioural changes, and "pathological ageing" (or "Alzheimer-type ageing") is contemplated. The study of non-human primate species is of prime importance for understanding AD but needs to be performed using precise, standardised protocols.

Introduction

When discussing Alzheimer's disease (AD) we generally think in terms of the mental and behavioural problems suffered by many elderly people (sporadic AD), or of those borne by younger adults who belong to a small number of families genetically afflicted by the disease (familial AD). The idea that AD may affect other species does not usually cross our minds. It is even common for specialised, learned literature to describe AD as a neurodegenerative disease specific to humans [1], a supposition that the limited number of studies undertaken on the possible existence of 'AD' in other species – even in non-human primates, our closest relatives – has generally upheld. However, sufficient data have now been published to warrant a re-think of this position.

The major problem with regard to affirming or negating the existence of AD in other species is the difficulty in interpreting the symptoms and signs presented by possibly affected animals. In theory, determining whether an animal is suffering from 'AD' requires examination to see whether the symptoms and signs that define the disease are all present. The WHO's definition of AD [2] states that Alzheimer's disease is a primary degenerative cerebral disease of unknown etiology, with characteristic neuropathological and neurochemical features. Clinically, patients suffer a form of dementia that is usually insidious in onset and develops slowly but steadily over a period of years; further, there is an absence of clinical evidence, or findings from special investigations, to suggest that the mental state may be due to other systemic or brain disease which can induce a dementia. Given our lack of knowledge regarding the aetiology of AD, its diagnosis places special emphasis on the simultaneous existence of two types of pathological change noted by Alois Alzheimer over 100 years ago: neuropsychological symptoms indicative of dementia for which complementary studies can find no cause - thus establishing a clinical in vivo diagnosis of AD via exclusion, and neuropathological signs indicative of neuronal degeneration and the accumulation of aberrant proteins (manifestations of cytomorphofunctional and cerebromolecular abnormalities) – thus establishing a post mortem anatomopathological diagnosis.

With respect to the mental and behavioural changes associated with AD, the WHO definition [2] states that dementia is a syndrome..., usually of a chronic or progressive nature, in which there is disturbance of multiple higher cortical functions, including memory, thinking, orientation, comprehension, calculation, learning capacity, language, and judgement associated with important changes in conduct (increased apathy, aggressiveness, etc.). Conduct changes have been least studied of this list, even though they may be the most important in the everyday relationship between patient and caregiver [3]. Certainly, if this definition of the mental/behavioural decline seen in AD is used strictly it can apply only to humans. Not even our closest evolutionary relatives could be said to show any significant deficits of this sort since they simply do not perform most of the cited higher functions. However, some of the above problems might certainly occur in senile animals, such as a diminished ability to learn or perform certain tasks, along with behavioural problems such as increased apathy or altered social relationships. Perhaps we need to consider more deeply the meaning of the disturbance of multiple higher cortical functions. If we conclude that a species has a certain number of such functions, then an individual of that species might be said to suffer a form of 'AD' if disturbances in these functions occur alongside cortical morphofunctional alterations similar to those seen in the brains of humans with AD (i.e., a marked reduction in the population of neurons, particularly in the hippocampus, substantia innominata, locus ceruleus, and temporoparietal and frontal cortex; appearance of neurofibrillary tangles made of paired helical filaments; neuritic (argentophilic) plaques..." along with gliosis, neuroinflammation, and oxidative stress etc. [3, 4-9]).

The definition of the higher cortical functions of each species, the recognition of their boundaries, and the identification of the neuronal circuits forming their neuroanatomical basis are not easy tasks. Neither is it easy to establish equivalences between the cerebral functions and neuronal circuits of different species [1, 10]. Thus, many studies on cognition in normal and AD animal models (via tests of learning, memory and attention involving systems in which animals must solve a dilemma or perform a task [1, 10]) have returned unexpected and even contradictory results. For example, causing a lesion in the basalocortical cholinergic nuclei that innervate the cerebral cortex may produce different alterations in the cortical functions of different species; certainly, the alterations seen in the performance of learning tasks and attention are different between monkeys and rats with induced lesions in the nucleus basalis. In most studies of this kind, it has not been possible to correlate the neuropathological alterations induced with any alteration in higher cerebral functions. In addition, animals with induced lesions in a brain area do not usually show the characteristics associated with senile degeneration in that area.

Attempts to determine whether 'AD' occurs in other species might do well to examine whether our closest relatives - the non-human primates - suffer any pathological process that might be termed AD. This chapter examines the literature referring to possible characteristics of AD in such primates, and some other animals, with special reference to the neuropathological features related to post mortem diagnosis.

Senile Involution in Non-Primate Mammals. The Scarcity of Ad Neuropathology

The literature contains few studies in non-primate mammals that describe the existence of extraneuronal amyloid and/or intraneuronal fibrillary abnormalities similar to those seen in human brains affected by AD, although a handful do mention them in dogs [11, 12], cats [13] bears [14] and (especially) wolverines (*Gulo gulo*) [15].

Nonetheless, no links with any deficits in the higher cerebral functions of these animals have ever established, nor have any comparisons been made between normal senile members of these species and those showing pathological signs reminiscent of human AD. Further, the majority of the most commonly studied species show no neuropathological abnormalities of this kind, although spontaneous senile alterations in behaviour are known to occur. Certainly, rats and mice have been followed over their natural lifespans without any amyloid accumulations or neurofibrillary tangles appearing spontaneously, even in strains or individuals showing significant deficits in learning abilities and memory. It is true, however, that other morphofunctional abnormalities have been recorded in these animals, such as an increase in gliosis and oxidative stress [16, 17]. In experimental rodent models, particularly rats with induced lesions in the basalocortical cholinergic centres, large deficits may appear (though certainly not always) in certain cortical functions, depending on the model sub-type [1, 10].

However, differences are nearly always seen between the neuropathological abnormalities induced and the behavioural deficits observed [1, 10]. Further, no signs of intra- nor extraneuronal fibrillary pathology are ever seen [1, 10], even though amyloid precursor protein (APP) synthesis increases following lesion induction [16] and intense neuronal abnormalities and pronounced gliosis may occur in the mid-long term [1, 17]. This apparently manifest inability to induce amyloid pathology might be due to rodents lacking an amyloid production system, the blockage of pro-amyloid factors, and/or some special resistance to the changes required for amyloid deposits to form.

Transgenic mice containing abnormal human APP and presenilin 1 (PS1) genes from patients with familial AD do form amyloid deposits similar to those seen in humans (since the defences against murine APP are breached), but not even these appear to provoke alterations in the higher cerebral functions [9, 18-20]. It was recently reported that rats subjected to certain treatments with amphetamines showed no neuropathological abnormalities but did show increased oxidative stress, greater immunoreactivity to anti-amyloid antibodies (APP, 6E11, AG8), and an increase in β-amyloid (detected by infrared microspectroscopy and Raman spectroscopy) in the cerebral cortex and hippocampus [21, 22].

The possible induction of apoptosis and the accumulation of amyloid substances in the rat hippocampus alongside the suffering of cognitive deficits have also been reported [23]. However, these animals, all of which were senile, were also deficient in vitamin E and kept under conditions of oxidative stress.

Finally, after inducing brain lesions, some subspecies of rabbit show cerebral neurodegeneration very similar to that seen in human brains affected by AD, although they do not produce amyloid [24, 25].

Thus, it appears generally true that most non-primates appear to show great resistance to the build-up of amyloid deposits in the brain, but that 'forcing' their physiological systems can induce certain AD-like neuropathological signs.

Senile Involution in Non-Human Primates. Ad Neuropathology and Mental/Behavioural Characteristics

All humans belong to the species *Homo sapiens sapiens*, and to the order *Primate*. Humans posses a complex brain able to perform higher functions [26, 27] that differ greatly from those performed by other primates. Over evolutionary time, natural selection favoured those species with brains of greater functional capacity, achieved through the so called process of '*encephalisation*' [28, 29].

The jumps between a lesser and the next more evolved primate are noticeable in terms of the development of different regions of the neocortex, the prefrontal cortex, and the centres and areas of association. Differences are also obvious in terms of the number of synaptic connections and the 'higher' cognitive and behavioural functions that can be performed [30-33]. The advances and morphofunctional perfections achieved in brain areas related to learning and the appearance of different types of memory etc., appear as 'jumps' between species of different evolutionary level.

However, language, the ability to reason and judge, and complex processes of learning and memory etc., are characteristics understood to be exclusively human [32-35]. Interestingly, the areas responsible for these abilities are those most affected by AD.

But do non-human primates also suffer from AD or something similar? In fact, only a few species have been studied in this respect. Table I shows those that have received the most attention within the suborders *Strepsirrhini* and *Haplorrhini*.

The latter suborder is divided into the *Platyrrhinii*, or New World monkeys, and *Catarrhini,* or Old World monkeys (which includes the apes). Humans fit into the latter category within the group of the large anthropoid apes, i.e., in the family *Hominoideae*, which contains the genera *Pongo*, *Gorilla*, *Pan and Homo*. There are 22 genera of prosimians, of which only two have been studied with respect to AD, and 49 genera of simians, including eight genera of anthropoids. However, within this latter, higher suborder, it is not the apes that have been most studied alongside humans with respect to AD, but the members of *Cercopithecus*, their evolutionary precursors. The following lines discuss some of the most important work undertaken in this respect.

Table I. Non-human primates that have received the most attention in the study of the normal and pathological senility within the suborders *Strepsirrhini* and *Haplorrhini*. The latter suborder is divided into the *Platyrrhinii*, or New World monkeys, and *Catarrhini*, or Old World monkeys (which includes the apes)

SUB-ORDER	INFRAORDER	FAMILY	GENUS	SPECIE
STREPSIEEHINI (PROSIMIANS)	LEMURIFORMES	CHEIROGALEIDAE	*Microcebus*	*M. Murinus* (Mouse lemur)
		LEMURIDAE	*Lemur*	*L. Catta* (Ring-tailed lemur)
		LEPILEMURIDAE		
		INDRIIDAE		
	CHIROMYIFORMES	DAUBENTONIIDAE		
	LORISIFORMES	LORISIDAE		
		GALAGIDAE		
HAPLORRHINI (SIMIANS)	TARSIIFORMES	TARSIIDAE	*Tarsius*	*T. syrichta* (Philippine tarsier)
	SIMIIFORMES- PLATYRRHINI (NEW WORLD MONKEYS)		*Cebus Callithrix* *Cebus Saginus*	*Diverse species:* Marmosets, Titis
		CEBIDAE		*S. Imperator:* Tamarins
			Saimiri	*S. Sciureus* (Squirrel monkey)
		AOTIDAE		
		PITHECIIDAE	*Callicebus*	*C. Moloch* (Red-bellied titi monkey)
		ATELLIDAE		
	SIMIIFORMES-CATARRHINI (OLD WORLD MONKEYS)	CERCOPITHECOIDAE	*Erythrocebus*	*E. Patas* (Patas monkey o Wadi monkey)
			Chlorocebus (*Cercopithecus*)	*Ch. Aethiops* (Vervet monkey)

Table I. (Continued)

SUB-ORDER	INFRAORDER	FAMILY	GENUS	SPECIE
			Macaca	*M. Mulatta* (Rhesus macaque)
				M. Fascicularis (Crab-eating macaque)
			Papio	*P. Cynocephalus* (Yellow baboon)
		HYLOBATIDAE		Gibbons
		HOMINIDAE	*Pongo*	*P. Pygmeus* (Orangutan)
			Gorilla	*G. Gorilla* (Western gorilla)
			Pan	*P. Troglodytes* (Chimpanzee)
			Homo	*H. Sapiens* (Human)

3.1. Physiological and Pathological Cerebral Involution in Non-Human Primates

Non-human primates are good animal model candidates for research into human senility and age-related diseases.

Their phylogenetic proximity means many of their brain areas show structural similarity to those of humans, with whom they also have certain cognitive processes in common (e.g., those related to visual, non-spatial abilities).

They can also perform certain tasks. However, only a small number of non-human primate species have ever been studied (Table I).

The reasons for this include the problems of raising such animals in captivity, the longevity of some species under such conditions, the small numbers found in the wild (which make them difficult to find and even harder to follow), and national and international legal restrictions regarding their use in scientific research.

Further, little work has been done comparing the cognitive/behavioural deficits and cellular/molecular alterations within and between species at different evolutionary levels.

This type of comparative work is essential for understanding both physiological ageing and the pathological characteristics of neurodegenerative disease.

Some senile non-human primates are reported to show (with greater or lesser intensity) both neuropathological abnormalities and changes in social behavioural, learning and memory [36-42], but only a relatively small number of individuals of just a few species have been studied behaviourally while alive and then anatomopathologically following their death [39].

In some of the few studies of this sort that do exist, very marked difference have been reported in social behaviour, the undertaking of learnt tasks, and the learning of new ones, between senile primates with no amyloid abnormalities and those of the same species with such abnormalities and showing notable neuronal loss [39].

However, these results cannot be considered confirmation of the existence of a universal dual model of senile involution (physiological, and senile AD-pathological) applicable to all primates; differences have been detected between elderly people and long-lived non-human primates that prevent the recognition of intercorrelations between behaviour and neuropathological abnormalities [40, 43, 44].

3.2. Life Expectancy in Non-Human Primates and the Passage from Maturity to Senility

Life expectancy varies a great deal between species and even subspecies, and depends strongly on both intrinsic and extrinsic factors. The lifespans of animals living in captivity or semi-captivity may be 30% longer than those of the same species living in the wild. In the latter, longevity depends on the healthiness of the habitat, the number of predators and the availability of food and water, while for those living in captivity or semi-captivity it depends on the quality of healthcare and food received, and on the characteristics of the habitat provided. In the USA, where experimental animals or those that have served some other purpose are 'retired' and looked after until they die, very old ages may be reached. For example, the Creative Habitats and Enrichment for Endangered Threatened Apes (C.H.E.E.T.A.H.) Primate Sanctuary in Palm Springs, California, cared for a chimpanzee that, at over 50, became the oldest in the world. That said, the variation in longevity of members of the sister genera *Macaca* and *Cynomolgus* raised at research centres, on reserves or in protected areas (for example *Macaca* arctoides on a number of small Caribbean islands), can be very pronounced [45, 46]. Even for the rhesus monkey *Macaca mulatta, Zimmerman*, 1780, singularly the most studied primate in all scientific settings, life expectancy and the moment of transition from maturity to old age remain matters of debate. In general, its life expectancy in protected habitats, captivity or semi-captivity is some 35-40 years, with those over 20 years of age regarded as old and those over 30 regarded as being of extreme old age [44-46]. It is interesting to note that despite the number of studies involving this species, not one offers a definition of physiological senility for it, nor any has any neuropathology (either morphological or behavioural) based on measurable characteristics been defined.

Some animals provide clear examples of the influence of the environment on lifespan and senile involution. Consider, for example, the *Chlorocebus* monkeys (*Chlorocebus* or *Cercopithecus aethiops*) of St. Kitts, an island in the eastern Caribbean where some 30,000 animals have been studied by the Behavioural Science Foundation (BSF) since 1968 [47-49]. The first of these monkeys to reach the island came from Africa on slave ships around the year 1630, and rapidly acclimatized. These omnivorous creatures live for some 20-30 years, as they also do in captivity, exceeding the life expectancy of their West African ancestors by some 5-10 years. The reason is simple: they are not exposed to the viral pathogens that decimate their African colonies [47]. This species may in fact be of great interest for the study of neurodegenerative diseases if its ageing process can be well characterised, since amyloid deposits are always accumulated [49].

In contrast to the prolonged lifespan enjoyed by macaques, which hinders the longitudinal study of disease in these animals, other species have much shorter lives, such as the mouse lemur *Microcebus murinus*, a small animal weighing just 100 g and which can be raised in small colonies in semi-captivity [42, 50]. However, the dependence of the ageing process on environmental factors in this species is made keenly aware: while in the wild these animals may live only 3-4 years, in captivity or semi-captivity they may reach the age of 10-13 [42, 50]. Such changes in their ageing process, and the very particular neuropathological changes that occur in their brains [43], render these animals controversial as a model for the study of human ageing and senility.

3.3 Physiological and Pathological Morphofunctional Cerebral Involution in Non-Human Primates

Background

Generally speaking, the morphofunctional alterations seen in normal senile human subjects [1, 4, 10, 51-55] are also seen in non-human primates. Table II shows the most studied features of what might be called "physiological senility", i.e., morphological and functional involution with mild and defined losses of certain cognitive abilities plus behavioural alterations. These phenomena are the consequence of neuronal dystrophy (alterations in the soma, dendrites and axons, loss of synaptic contacts) and gliosis [36, 55-72]. The majority of work investigating the existence of neurodegenerative processes related to AD has focused on the search for neuropathological abnormalities similar to amyloid plaques and neurofibrillary tangles [73-92]. Very few studies have examined any other morphological or biochemical variables such as synaptic [93] or oxidative stress markers [94]. Table III shows the most important results published to date. Accumulations of β--amyloid similar to plaques, with shapes and immunoreactivities both similar to and different from those described in humans, as well as diffuse, non-vascular deposits, can be found in senile individuals of nearly all non-human primate species, the load increasing with age. However, only some macaque and chimpanzee brains show degenerative changes in neuronal bodies and neurite dystrophy near these deposits (mostly in the hindbrain) [39, 40, 74, 76, 80], with some animals showing reductions in the function of neurotransmitter systems involved in human AD, particularly the cholinergic system [74, 91].

Table II. Morphological alterations in "physiological senility" of non-human primates [55-72]

Loss of cortical neurons (brain cortex and hippocampus)
There are quite a few studies with very contradictory results: from significant neuronal loss [36, 55] to the preservation of the number of neurons [56-59], even in animals with some cognitive deficits [60]. The most affected regions are the prefrontal cortex and CA1 area of the hippocampus [55]. The involution of some areas may correlate with cognitive senile involution, but not with others [61]
Loss of subcortical neurons
There are very few studies. Significant neuronal loss (except in one work [68]) in cholinergic basalocortical nuclei (septum and nucleus Basalis of Meynert – nbM [69, 71]) and brainstem nuclei (reticular systems, locus coeruleus and Raphe dorsalis [69, 70].
Changes in neuronal bodies, axons, dendrites and synapses
Vacuoles and deposits in cortical pyramidal neurons but not in hippocampus [59, 62]
Hypertrophy of some cholinergic neurons in the brainstem [69, 71]
Major changes are observed in dendrites, axons and synapses in cortex (prefrontal, temporal, visual cortex) and hippocampus [62-67]. The most important alterations are considered as a first step for neuropathological alterations [65].
Alterations of the myelination
Demyelination in axons of cortical neurons and decrease in the volume of white matter [58, 71, 72]

The age of onset of amyloid neuropathology, the type of deposit laid down (perivascular or parenchymal), where this occurs, and its relationship with other neuropathological signs varies considerably between species, and maybe even between individuals of the same species given the very different descriptions made by different authors. For example, amyloid deposits are not seen in proprimates of the order *Scandentia*, such as *Tupaia belangeri* (the tree shrew of Southeast Asia) [41]. The shape of accumulations also varies widely between species, but follows no trend that might be associated with evolutionary level. Perivascular deposits are particularly abundant in prosimians and certain lower New World monkeys, and may or may not be associated with other neuropathological signs depending on the species and the age of the animal [43, 84]. Finally, the rhesus monkey shows neuritic plaques, mainly in the parenchyma of the cerebral cortex, while in the squirrel monkey perivascular deposits are more abundant, and in chimpanzees mostly diffuse plaques are seen [76-78, 84]. In all cases, including that of Tupaia, it has been shown that the peptide sequence of β-amyloid is identical to that of humans. Two studies have shown that in non-human primates there is a single isoform of ApoE associated with amyloid deposits, which is analogous to human ApoE4 [77, 78]. Intraneuronal fibrillary tangles, one of the hallmarks of AD, are not usually seen in non-human primates. Two species, however, do show apparent tauopathy with high concentrations of phosphorylated and non-phosphorylated tau protein: the more lowly *Microcebus murinus* [42-43] and the much higher *Chlorocebus aethiops* [49] (note: more recent studies – Table III- discussed below have detected tau abnormalities in Old World monkeys, though unlike those seen in humans). In general, dystrophic dendrites immunopositive for phosphorylated tau are found in animals showing the greatest involutive abnormalities and greatest amyloid pathology, and show an intimate relationship with amyloid deposition in the majority of brain areas [87, 88]. The protein metabolism pathways that give rise to β-amyloid deposits and neurofibrils have only been partially studied in monkeys, although they appear to be the same as those seen in humans [79, 88, 93, 94].

Cerebral Involution in Lemurs (Prosimians)

The most studied lemur model of AD is that of the mouse lemur *Microcebus murinus* [42-44, 50, 75, 79, 95, 96]. Similar to a small rodent, this species has the experimental advantages of being small and having a short lifespan – animals over five years of age are considered very old. The majority of lemurs show slow, progressive cerebral ageing without much involution, and some 20% show early neurodegeneration.

Table III. Neuropathological abnormalities ("Alzheimer's type") in senile non-human primates [36-40, 73-92]

Non-vascular accumulations of β-amyloid.
Amiloyd deposits in senile individuals of all species of non-human primates have been described [36-40, 45, 47, 49, 50, 65, 73-83], with greater or lesser density and different aspects and locations depending on the species.
A β-40 is the predominant amyloid peptide [39] and ApoE4 can be associated in the deposits [77, 78].
Relationship with vascular deposits is very variable [81, 82, 84].
Amyloid burden increases with age, mainly in frontal and parietal lobe. Low density of deposits are observed in occipital lobe, prefrontal cortex and parahippocampal regions [77-83].
There is no relationship between number of plaques and cognitive damage [74, 76].
Amyloidosis can be induced experimentally [82]
Neuronal and axonal loss are selectively linked to fibrillar amyloid within plaques [90]. Some macaque and chimpanzee brains show degenerative changes in neuronal bodies and neurite dystrophy near these deposits (mostly in the hindbrain) [39, 40, 74, 76, 80]
Perivascular amyloid
Deposits on blood vessels are very marked in many prosimian species and lower New World monkeys [81-84].
Different level of accumulations are observed in more evolved simian species [76-78, 84, 85].
Tau pathology (immunopositive intraneuronal deposits for non-phosphorylated and phosphorylated Tau)
Not often seen neurofibrillary tangles (NTF-like) in non-human primates [85-88] except in prosimians (*Microcebus murinus* [42,43]) and *Chlorocebus aethiops* [49].
Tau accumulation has become apparent in recent years in in Old World monkeys, though unlike those seen in humans, (*Pan Troglodithes* [86], *Macaca fascicularis* [87, 88], *Gorilla gorilla* [89])
Dystrophic neurites, Tau-inmunopositive, in some individuals of different species
Colinergic involution
Involution of neurotransmitter systems involved in human AD, particularly the cholinergic system [74, 91, 92].

In these primates, neurodegeneration is characterised by massive atrophy of the cortex, hippocampus, basal ganglia, brainstem and cerebellum, associated with a considerable increase in the size of the ventricles, abundant amyloid plaques, accumulations of tau protein and the loss of cholinergic neurons.

As well as amyloid plaques, other types of amyloid deposits develop, along with argentophilic neuronal filaments like those seen in many patients with AD. Animals that suffer this degeneration lose cognitive/behavioural abilities (the ability to perform tasks, abnormal social relationships) resembling to some extent that seen in human AD sufferers.

Studies on the composition, development and distribution of amyloid plaques in *Microcebus murinus* have also highlighted certain similarities with humans. In *Microcebus murinus*, β-amyloid is deposited as diffuse plaques about 100 μm in diameter, as dense plaques with a compact amyloid nucleus (thioflavin S-positive), and as belt-like formations around blood vessels.

The first two types are also found in the brains of patients with AD. As in humans, the amyloid deposits seen in *Microcebus murinus* have two main constituents: A β-42 and Aβ-40. The deposits around cortical blood vessels are immunopositive for both, while the diffuse plaques are highly positive for Aβ--42, with some containing only Aβ-42. This suggests that, as in humans, Aβ-42 is associated with early plaque development [97].

Although the Aβ-40/Aβ-42 ratio is greater in *Microcebus murinus* than in humans, suggesting small differences in the corresponding amyloid production mechanisms [97], the distribution of plaques is similar and involves the upper part of the temporal lobe, the amygdala, the prefrontal area, the parietal lobe, and the occipital lobe (in descending order of density in *Microcebus murinus*) [43, 44]. Sequencing of the *Microcebus murinus* gene for APP has also shown that despite differences in some nucleotides, both its APP and Aβ-peptide are completely homologous to the human forms [75].

Microcebus murinus APP localises in cell bodies and the proximal dendrites of neurons, in astrocytes and oligodendrocytes, in amyloid deposits of the cortical parenchyma, and on the walls of blood vessels, similar to that seen in humans. In addition, a good correlation has been reported between APP levels and the density of amyloid plaques, with both increasing as animals age [43, 75].

The accumulation of tau protein is also seen in *Microcebus murinus*, and increases with age. The area most affected is the neocortex in young animals, while the subiculum and entorrhinal cortex are also affected in eight year-old animals (i.e., those of very advanced age); this, however, is quite different to that seen in humans. While all animals that show diffuse β-amyloidosis all have accumulations of tau in the neocortex, there appears to be no relationship between their relative densities [43, 75].

Cerebral Involution in Cercopithecans (Old World Monkeys)

The genera *Chlorocebus* and *Macaca* have attracted most attention in studies on cerebral involution. The *Chlorocebus* monkey *Chlorocebus* aethiops (in particular its colonies on the island of St. Kitts) has recently been proposed as a model for studying AD amyloid pathology [47-49]. Adult males may weigh some 5-7 kg and live 20-30 years in captivity (see Section 3.1). Studies performed on individuals aged 15, 22 and 30 years shown that amyloid deposits (mostly containing Aβ-40) begin to appear at about the age of 15 years, localised around the blood vessels. No plaques are seen in the cerebral parenchyma at this point.

All 22 and (with much more intensity) 30 year-old animals show plaques in the frontal, temporal and occipital cortices, in which Aβ-42 is much more common than Aβ-40 (as seen in human brains affected by AD). Fibrillar amyloid, detected by thioflavin S staining, appears around the blood vessels at all adult ages (although more at the occipital level, unlike in humans), but only a small proportion of this forms parenchymal plaques in animals aged 22 and 30 years. At 15 years of age there appear to be no nueronal or glial abnormalities, and only when animals reach 22 years of age are some neuritic-like plaques seen.

However, all 30 year-olds show a wide range of pathological abnormalities typically associated with amyloid plaques in humans (reactive astrocytes, activated microglia, dystrophic neurites). The great quantity of neuritic plaques seen at this age, which are of a density similar to that seen in human AD, contrasts with the scarcity of these structures in the majority of non-human primates studied to date [47-49].

Although these plaques are immunopositive for several amyloid substances, including APP, only a very small percentage are positive for the phosphorylated tau characteristic of human dystrophic neurites. No neurofibrillary tangles immunopositive for phosphorylated tau are seen.

The rhesus monkey *Macaca mulatta* and the long-tailed or crab-eating macaque or *cynomolgus* monkey *Macaca fascicularis* (or *Macaca cynomolgus*) are the most commonly used non-human primates in brain research. Young animals (up to 5 years of age) of these species show no cerebral amyloid deposits, but nearly all old animals (25-30 years of age) have deposits in the cerebral cortex [88, 94, 98], although the degree of β-amyloidosis and its functional implications are the subject of much discussion (see Section 4.7).

The incidence of intraneuronal neurofibrillary pathology is generally regarded as minimal or zero but in the last years a degree of Alzheimer-type tau pathology has been observed [87]. The APP of these animals shows great homology with that of humans. Many more of their cortical plaques are immunoreactive to Aβ-40 than to Aβ-42. About 50% of these plaques contain acetylcholine esterase of the same biochemical characteristics as the human enzyme.

In some 20% ApoE is detected, and many plaques contain small quantities of proteoglycans, heparin sulphate and β-chymotrypsin. These characteristics are all seen in humans, although the distribution of plaques and their progression in space and time is different. For example, in humans the hippocampus, amygdala, entorrhinal cortex and the frontal and parietal lobes are more affected, and the classic spatiotemporal progression of the neuropathological alterations is not seen in these species.

Animals with moderate Aβ loads show deposits in the cortical association areas, those with higher loads also show deposits in the paralimbic areas, and those with high loads have plaques in the limbic areas (although the shape can be very different). Some authors report a greater concentration of plaques in the frontal areas and primary somatosensory cortex, and lesser concentrations in the amygdala, insula, cingulate cortex, the temporal limbic region, the occipital cortex, and parietally-associated cortices [94]. The motor cortex and some sensory areas do not seem to develop deposits.

The stump-tailed macaque also named bear macaque (*Macaca arctoides*), I.G. Saint-Hilarie, 1830, is native from Southern Asia and it is distributed from India, Burma, China, Bangladesh, Malaysia and Thailand, where it lives at altitudes of up to 2000 m, according to rain forest distributions and food availability [98]. There is a definite sexual dimorphism (males with highest size and robustness), measuring 45-70 cm in height and weighting 7-12 kg [99]. Like other *Macaca* species, it has been successfully introduced at certain Caribbean Islands and cays, and has been bred in captivity and semi-captivity by many Central American laboratories [99].

Stump-tailed macaques may live for more than 30 years when kept in a protected environment. They exhibit age-related morpho-functional changes particularly in the Central Nervous System, similar to that seen in the rhesus monkey, although with some notable differences [36-38]. During more than 15 years, the Cuban Brain Aging and Alzheimer's disease Basic Project has collected behavioural (social patterns); neuropsychological (reaction time, recognition memory and executive function tests); neurophysiological (quantitative electroencephalography, somatosensory and visual evoked potentials) and imagenological (MRN-2T and CT scan) data from more than 30 stump-tailed macaques (7 to 34 years of age) [36-38, 94, 100, 101]meanwhile the Morphology laboratory at Cajal Institute, Spain has been carried out brain morphological studies. Delayed matching and nonmatching-to-sample (DMS and DNMS) tests have proven to be very useful for determining the differential degree of learning and memory abilities in monkeys [102]. In humans, the DNMS task also has been shown to be sensitive to selected temporal lobe damage [103], and to Alzheimer's disease and normal aging [104]. In general, our macaque group exhibited a decline in objective cognitive ability and behaviour no before 20 years of age. Similarly to that observed in mature-old rats and dogs [105, 106] aged monkeys -as a whole group- show a significant impairment compared to young controls in the above cited tasks but the use of exhaustive statistical analysis (cluster analysis), allowed us to detect "impairment subgroups" considering the two variables contributing mostly to each load factor (latency and errors). Cluster analysis is a special statistical procedure that measures the degree of similarity among individuals based on the relative "distance" of the data.

Fixing the Euclidean distance lower than 20, this test revealed the separation of three clusters, that is, non-impaired, middle- and severe impaired monkeys (Table IV) similarly to that observed in other human [107] and nonhuman primates [108]. Brain post mortem examination of severely impaired monkeys (aged 25-34 years) showed significantly more notable pathological abnormalities related to the accumulation of amyloid than seen in animals with mild behavioural deficits. High plaque densities were seen in prefrontal cortex, hippocampus and parahippocampal regions, with a distinctive polymorphism in the plaque type (Figs. 1 and 2).

Oddly enough, some areas in the vicinity of these regions showed no such accumulations, suggesting that neurodegeneration is not a general process and may be localized, associated or not to particular neurotransmitter circuits (P-R). Other brain regions showed disperse plaques. Diffuse immunoreactive deposits were also seen in all regions of the neocortex formed by groupings of small areas showing homogenous immunostaining or very fine immunoreactive extracellular and some intraneuronal granules (Figure 2).

All these deposits were visualised with human anti-Aβ-40 and/or anti-Aβ-42, although Aβ-40 was the main component, in agreement with that described for higher simians. In some animals, small intraneuronal accumulations of protein immunoreactive to human anti-tau and anti-phosphorylated tau antibodies were seen (Figure 2). It is a relevant finding since tau pathology was once believed exclusive to humans and mouse lemurs, although tau-like neuropathology has been now reported in Chlorocebus and in some individuals of Old World non-human primate species (*cynomolgus*, *chimpazee*, *gorilla*) [49, 86-90] (Table III). In these most impaired individuals other neuropatological alterations are better observed such as astrogliosis and microgliosis, neuronal inclusions, dendrite and synapse loss and major abnormalities (Figure 2).

Table IV. Mean Performance Scores for Young and Old Monkeys in the DNMS TASK (See Text)

Parameters		Age Classification		Aged group classification (Cluster analysis)		
		Young Group	Aged Group	Old non impaired	Old-mild impaired	Old-severely impaired
Initial Learning	Trial(*)	250	867.5	370	570	915
	Errors(**)	25.5	304.5	50	140.5	295.5
Retention		96.6	76.0	88.0	85	80.5
Sequential Delay testing	30 s	93.5	80.25	90.0	85.5	76.0
	120 s	89.0	73.0	87.5	80.5	72.5
	360 s	81.5	66.75	84.5	74.0	69.0

(*) Number of trials and errors required to reach the learning criteria.
(**) Scores expressed as percent of correct choices.

Alongside these animals with AD-like neuropathological abnormalities, others with much less serious cognitive/behavioural problems and scant neuropathological alterations have been recognised. In both types, the cellular density is very similar to that seen in animals under 20 years of age. No intense gliosis appears to occur – unlike that seen in humans - although one animal among all those studied did show such signs.

Different species of macaque are now being examined for biomarkers that might help diagnose brain abnormalities. An increase in markers of oxidative stress has been found in peripheral blood [94], and our laboratory is studying the potential of the increase in β-configured proteins [109].

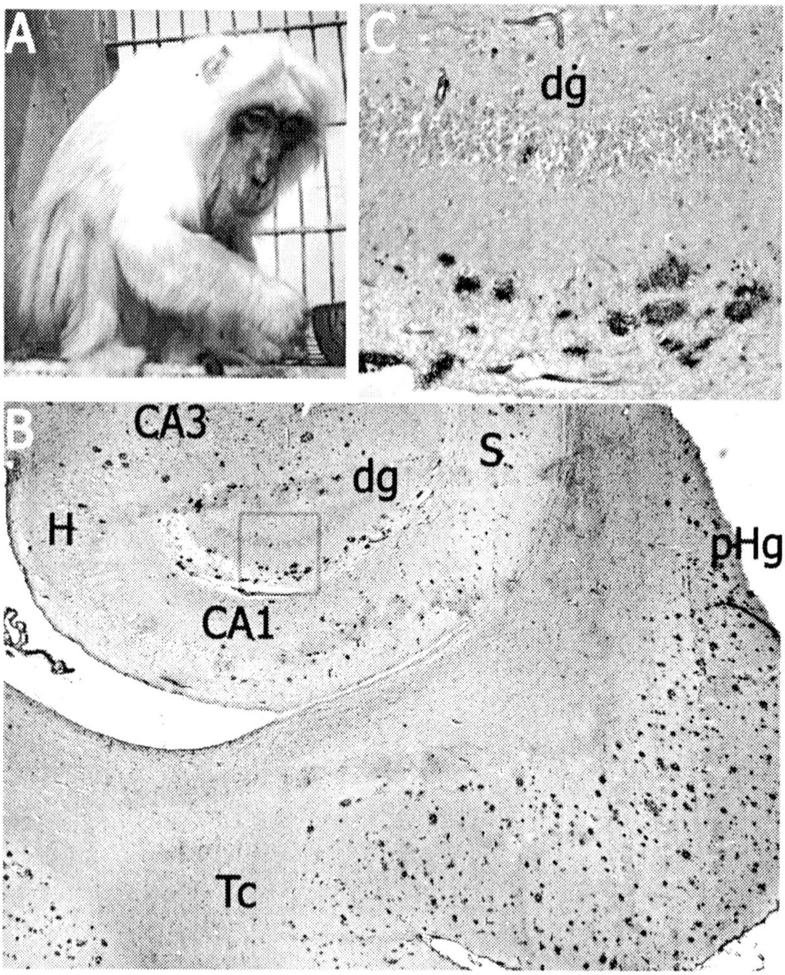

Figure 1. A. 34 year-old, female stump-tailed macaque (*Macaca arctoides*). Note the animal's grey hair. Two year ago this animal began to receive significantly lower scores in the DNMS-task test compared to other animals of the same age. Anatomopathological examination (post mortem) showed large accumulations of amyloid throughout the brain cortex and hippocampus. B. A very high density of different types of amyloid plaques (immunostained with antibody 6E10 – Signet/Covance, New Jersey USA- which recognises epitope 1-16 of human β-amyloid) in the hippocampus. In (C) are observed at higher magnification the plaques in the stratum moleculare of the dentate gyrus. CA1 and CA3 = fields of Ammon's horn (hippocampus). dg = dentate gyrus (hippocampus). H = hippocampus. pHg = parahippocampal gyrus. S = subiculum. Tc = temporal cortex.

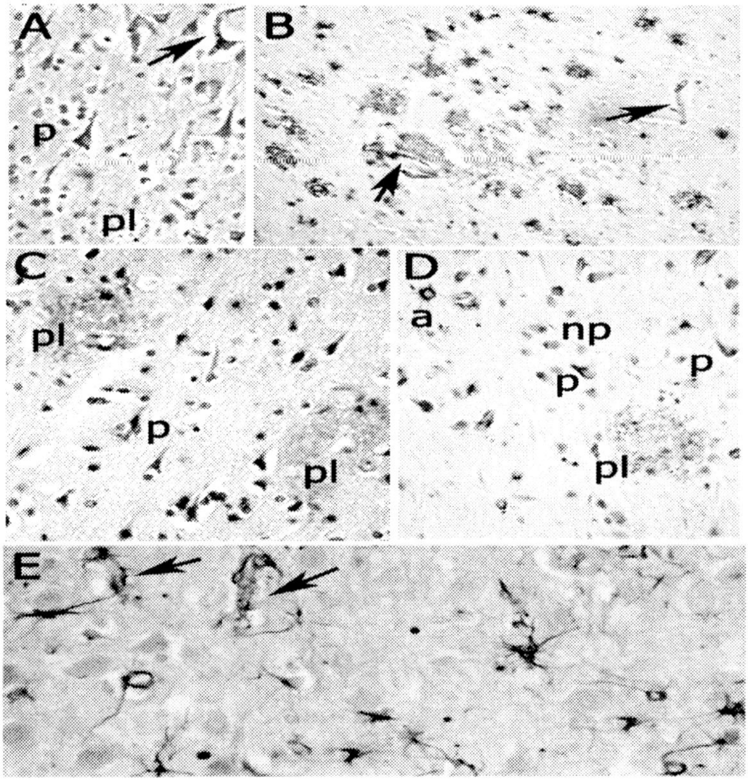

Figure 2. AD-like neuropathological alterations in the frontal cortex of the stump-tailed macaque of figure 1. A. Bielchowsky stain of layer V. Argyrophilic grain inclusions in pyramidal bodies (p) (tau inclusion-like), on blood vessels (arrow) (perivascular amyloid) and in neuropil (plaque amyloid) (pl). B. Different types of amyloid plaque (diffuse, neuritic-like, star-like) and perivascular amyloid (arrows) in layers III-V (immunostained with antibody 6E10 - Signet/Covance, New Jersey, USA - which recognises epitope 1-16 of human β-amyloid). C. Plaques (pl) of different type (immunostained with antibody 4G8 - Signet/Covance, New Jersey, USA -, which recognises epitope 17-24 of human β-amyloid) in layer V. Some pyramidal neurons (p) displayed granular amyloid deposits in its bodies. D. Pyramidal (p) and non-pyramidal cells of layer V showing tau accumulation (not seen in control animals of the same age), immunostained with antibody tau-2 (Chemicon/Millipore, Billerica, MA, USA), which recognises human tau protein, are observed. Some astrocytes (a) are also heavily immunostained (body, main processes, end-feet on blood vessels). Granular tau deposits are observed inside of some plaques (pl) (neuritic-like plaques). E. Hypertrophic/hyperactive astrocytes of different shape immunostained with antibody against GFAP (glial fibrillary acidic protein, Chemicon/Millipore, Billerica, MA, USA) in layer V (arrows, hypertrophic end-feet).

3.4 Mental/Behavioural Abnormalities in Senile Non-Human Primates

Normal ageing in humans is accompanied by the gradual loss of a wide range of mental capabilities. However, the deterioration of these functions differs from one person to another. The same appears to occur in many non-human primates, as determined by a number of important studies on memory, learning, cognitive flexibility (the ability to resolve a problem using different strategies), attention and the regulation of motor capacity. In addition, just like in humans, improvements can be made if certain types of training are undertaken [53, 54, 113]. Similar to that seen in humans, senile non-human primates take longer to learn new tasks and develop new abilities. This has been confirmed in normal animals and in those with small induced brain lesions via the use of behavioural tests such as the delayed response test (in which animals have to memorise a spatial location or recognise objects) and via cognitive flexibility tests [39, 40, 110, 111]. Both short-term and long-term memory are also affected. Numerous studies on aged monkeys, especially macaques, have shown that they face increasing difficulties in remembering different types of information [39, 44, 111-115], and many results are similar to those recorded for humans in the same tests, for example when subjects have to remember different places in a house illuminated one after the other [115]. The results of these studies also reinforce the idea that memory improves with certain types of training [113, 116]. Sometimes, however, clear differences are seen between humans and non-human primates. Many humans show important losses of attention with age, while in non-human primates this problem, although reported in some circles, does not appear to be of the same characteristics. For example, in studies on the speed of response to a stimulus, no differences were seen between young and old rhesus monkeys [39, 40]. Although the above reductions in brain function are well documented, no non-human primate equivalent of dementia has ever been reported since dementia is not a reduction in one or two cognitive functions but the sum of all the serious abnormalities and aberrations in all higher cognitive functions.

3.5 What Can Non-Human Primates Teach us About Ad?

The present section focuses attention on three areas that could increase our understanding of AD: a) the similarities and differences between AD-type neuropathological lesions in humans and non-human primates, b) the prevalence of these lesions and of a possible 'Alzheimer's syndrome' in affected animals, and c) the relationships between physiological senility and pathological senile involution (AD).

Similarities and Differences between AD-Type Neuropathological Lesions in Senile Humans and Non-Human Primates

In general terms it can be said that during senility primates develop marked neuropathology related to the accumulation of amyloid (although this is not true of all species and different individuals are affected to very different degrees), while only humans (the most highly evolved primate) and mouse lemurs *Microcebus murinus* (a lower primate) develop tau-related neuropathology. However, studies on lower and higher non-human primates have not only failed to clarify why amyloid and tau pathologies develop, they have produced more questions than answers regarding neurodegeneration in AD. Why β-amyloidosis is always seen in senility in some species (*cercopithicans*) and not in others (certain types of tamarind [41]), and why it should be associated with neurodegeneration in certain members of macaque species - as seen in human AD - is not easy to understand. It is also hard to understand the lack of a clear tau pathology in most primates.

The reason why amyloidosis occurs in primates remains unclear. APP is an ancestral membrane protein seen in unicellular organisms but expressed abundantly in mammalian brains [117]. It is well conserved, and in all mammals it appears to be broken down via two possible pathways: the amyloidogenic and non-amyloidogenic pathways [4, 6, 7]. However, β-amyloidosis is not seen outside of primates (except for in very rare cases in dogs, bears and wolverines etc.; see Section 2). In the tree shrew *Tupaia belangeri*, a pro-primate of the order Scandentia (much lower than the order Primate and in fact quite close to rodents), the sequence of the amyloid peptide is identical to that of humans, yet the former species suffers no kind of β-amyloidosis, either interneuronal or associated with blood vessels [41]. Members of the order Primate, however, clearly and consistently show signs of amyloid neuropathology related to old age, yet no correlation exists between the incidence, prevalence or intensity of β-amyloidosis experienced by each species and its evolutionary level. Neither is any correlation seen with cognitive/behavioural abnormalities. In those species and individuals in which amyloid does not accumulate, the elimination of insoluble β-configuration products may be facilitated or the breakdown products of APP may be different. With age, or in individuals with alterations in the factors that regulate amyloid production and/or its elimination [4, 6, 9], amyloid might accumulate. In humans the factors that might alter amyloid accumulation and/or elimination and lead to AD are oxidative stress, neuroinflammatory reactions, the induction of apoptosis and the entry of neurons into the cell cycle.

The morphological and morphohistochemical amyloid lesions seen in non-human primates are much more varied than those described in humans [80-83, 88, 97]. For example, lemurs and macaques have different types of deposit that can group into parenchymal plaques, diffuse parenchymal deposits (not in plaques), and perivascular deposits [43, 49, 88, 94, 98]. In human brains all these types are found [118], but non-human primates have more subtypes. This is hard to explain why these deposits showing molecular heterogeneity [119]. Vascular deposits appear to be very abundant in lower monkeys, forming belts around the vessels, but are much less abundant in anthropoids such as humans, in which it is only prominent in rare pathologies such as hemorrhagic amyloidosis [120]. Accumulations appearing as plaques in non-human primates may show more shapes and have wider immunoreactivities (e.g., against Aβ-40 and/or Aβ-42, and perhaps against neprilisin [121]) than those seen in humans. The diffuse amyloid deposits seen in non-human primates correlate much better with those seen in humans, both in the cortex and sub-cortical areas. Finally, the spatiotemporal development of amyloid pathology in humans is not seen in non-human primates; neither are the most affected areas of the brain the same.

The fact that tau-related neuropathology is not normally seen in non-human primates (Figure 2).) is one of the big constraints associated with contemplating the existence of "AD" in these species, and indeed in their use as models of the human disease. In both prosimians and simians neurofibrillary tangles and dystrophic neurons are practically inexistent, and there is very little immunoreactivity to tau (especially its highly phosphorylated form). Humans and great apes have very similar (99-100%) tau protein sequences and their differential susceptibility to develop filamentous tau inclusions may be influenced by intronic Tau sequences [122]. It might therefore be that the β-amyloidosis suffered by non-human primates is another disease(s), different to AD. These differences in the type of amyloid deposits and in tauopathy might be considered great inconveniences when arguing that non-human primates suffer AD. However, in some cases the amyloid load has been found to correlate well with cerebral atrophy [123,

124], with variations in cognitive functions/behaviour, and with alterations of tau metabolism (personal observations). This would appear to indicate that, although there are clear differences between the neuropathologies suffered by humans and non-human primates, they may be related.

Difficulties in Determining the Prevalence of "Alzheimer Syndrome" in Different Types of Non-Human Primate

Some non-human primate species show amyloid deposits quite consistently. The best example is the chlorocebus monkey *Chlorocebus aethiops*; several authors report 100% of senile animals to show amyloid deposits [49]. Although there have been few in-depth cognitive/behavioural studies on this species, it would appear that β-amyloidosis is not accompanied by serious alterations of brain function that might be interpreted the result of cerebral neurodegeneration. Macaques, which are evolutionarily close to *Chlorocebus* (indeed, they both belong to the family *Cercopithecoidea*) show a wider range of situations in senility. For example, the rhesus monkey (*Macaca mulatta*) shows a generalised, mild to moderate involution of the brain, with some individuals showing morphological and functional abnormalities more serious than those seen in the rest of the aged population [53, 54]. These individuals might be said to suffer a kind of Alzheimer's syndrome. However, we do not know (not even approximately) the percentage of such affected individuals in populations of this species since most studies have involved very few animals. There is one large wild colony being studied (by the Caribbean Primate Research Center of the University of Puerto Rico), that of Cayo Santiago just off the coast of Puerto Rico, which has over 1000 members distributed among several subcolonial families [46, 99].

However, these animals are not the subject of research into cerebral neurodegeneration/involution. Further, the information available on the symptoms and pathological signs in rhesus monkeys are very contradictory. Initially, authors who had studied AD found very serious anomalies in senile animals, such as intense losses of neurons and synapses and β-amyloidosis in the prefrontal cortex and hippocampus [55, 64, 73, 88, 94, 98].

Later studies showed that their morphohistochemical alterations were not so marked, and even reported neuronal losses to be scant, although important cognitive deficits were seen [60, 114, 125]. Indeed, very variable percentages of deficits have been recorded depending on the cognitive/behavioural tests [95] as well as individual differences across the adult life span [126, 127].

Nonetheless, the majority of authors concur in that only a few senile macaques show marked amyloid deposits plus serious cognitive/behavioural problems. In a very small number of animals neuropathological signs were recorded in the nucleus basalis [40, 91], just as in AD. Studies undertaken on the cynomolgus monkey Macaca fascicularis have returned similar results [87, 88].

An intense AD-like amyloid pathology appears in a variable proportion of senile animals, although perhaps no greater than 20-30% (it is hard to know for sure, however, given the small number of animals studied) [94, 98, personal observations].

However, the majority of animals that do develop these problems also show serious cognitive/behavioural deficits. Our laboratory's work on Macaca arctoides has shown that about 30% of senile animals (over 25 years of age) show intense AD-like amyloid pathology in several cortical areas, and they score poorly in different behavioural tests. The prevalence of this syndrome cannot be stated precisely, however, given the reduced number of animals studied [17, 18, personal observations]. 'Normal' senile animals, both in terms of their brain structures and behaviour, have also been recorded [110], this is the situation in which Moss and cols. have defined as "successful aging" [128].

Another important aspect that must be clarified is whether amyloid is actually neurotoxic in non-human primates. Accumulations of amyloid appear to be less neurotoxic to a species when they appear with high incidence and prevalence in senile animals, such as in *Chlorocebus aethiops*, for which no serious cognitive/behavioural problems have been recorded.

The difference in the neurotoxic properties of β-amyloid deposits between species may be due to them being of different composition and molecular organisation. When monomers and oligomers of β-amyloid aggregate to form fibrils and insoluble deposits (a process that is still not well understood), different types of physiopathological accumulation may occur, with some aetiopathogenic for AD in humans.

This idea is supported by the fact that some substances such as benzothiazole imaging agent Pittsburgh Compound B (used to mark plaques in single photon emission computed tomography studies) show different reactivity to human and non-human amyloid deposits; human deposits are marked by this compound but those of other primates (squirrel monkey, macaques) are not [129].

Physiological and Pathological (Alzheimer's) Senile Involution; the Continuum between Normal Senility and AD

The concepts that some authors have of AD are almost diametrically opposed. Some support the idea that sporadic AD is a neurodegenerative disease linked to ageing developed by certain genetically pre-disposed persons through the action of both intrinsic factors such as diabetes, and vascular problems, etc., and little known extrinsic factors – it is thus quite different to normal senility, 2) while others understand AD to be simply the last phase of senile cerebral involution. In fact, Alois Alzheimer himself even contemplated that what he had described might be a manifestation of intense, accelerated ageing of the brain. Indeed, many modern epidemiological studies have calculated that at an age of 100 years AD would be 100% prevalent, with the genes responsible for the most evolved regions of the human brain also being those responsible for the appearance of the "disease" [129-131]. It has been proposed that a continuum exists between normal physiological senility and the last stages of AD, passing through a stage of mild cognitive impairment (MCI), while yet others believe involution may be 'individualised' and have different outcomes [132, 133]. The involution of cerebral function in each individual would thus have its own characteristics, and like AD and MCI would appear at a variable age between 60 and 100 years, and with lesion intensity and neuropathological manifestations that do not correlate with age [132, 133]. Thus, different populations of different age might exist, composed of persons who are normal, normal with mild neuropathological abnormalities, with MCI, or with AD at different stages of clinical and anatomopathological decline. The results of studies on the prevalence of AD with age, the idea that some people with MCI go on to develop AD within 4-5 years [132, 134], and the characteristic spatiotemporal and clinico-pathological development of AD, are, for some authors, conclusive proof of a "continuum" between physiological senility and AD. However, these characteristics may only be indicative of the progression of AD and the apparent greater ease with which it occurs with increasing age. The subpopulation of elderly humans with no dementia, even those with mild-moderate neuropathological abnormalities, might be equivalent to the large subpopulation of cercopithicans with scant behavioural deficits. This subpopulation is of great theoretical interest since it might reflect the possible existence in senile physiological involution of a stable phase without dementia. Multidisciplinary research is required into how to stabilise senile involution in humans at this stage. The ultimate cause of AD in human brains must be sought in the genetic changes that have occurred over our evolution [129-131, 135-137]. Adaptation to ancestral environments when lifespans were shorter would have led to the appearance of an optimum genome, but with our current longer life expectancy these genetic changes might now be the source of serious pathological abnormalities at the end of life. The evolutionary step that gave rise to the primates appears to have also given rise to the appearance of a disease linked to ageing that is seen in its fullest form in humans. The different neuropathological manifestations of this, such as amyloidosis, tau-dependent abnormalities and oxidative stress etc., in different primates (including humans) are therefore a reflection of the evolution of the APP and tau genes, those coding for enzymes that regulate their metabolism, and those coding for the production and control of the free radicals responsible for oxidative stress [122, 131, 135-138].

Conclusion

β-amyloid-related neuropathological abnormalities similar to those that occur in human AD are seen at all evolutionary levels within the primates, although the pattern of amyloid accumulation differs. In contrast, tau-related abnormalities are not usually seen outside humans and mouse lemurs (low prosimians). Neither are reductions in cognitive/behavioural functions in non-human primates very well documented; although they appear to occur in all senile animals, only in some species are they manifested with any intensity. Analysis of the literature and of our laboratory's own observations in stump-tailed macaques suggests that in some non-human primate species, certain individuals suffer very notable neurodegeneration similar to AD while others experience senile involution with few neuropathological problems or cognitive/behavioural deficits.

However, this conclusion is certainly not applicable to all non-human primates, even to those that are evolutionarily close to *Macaca*, such as *Chlorocebus aethiops*. While members of the latter species consistently suffer neuropathological abnormalities they appear to experience few cognitive deficits. Despite the long, and variable lifespans of some non-human primates species, the lack of a definition for the onset of senility, and the unclear differences between normal senility and pathological changes (AD-like problems), the study of these species has allowed important conclusions with relevance to human AD to be drawn.

Acknowledgments

The authors thank Adrian Burton for language assistance and critical reading of the manuscript. Part of the work has been supported by a grant from the Spanish Ministry of Science (PN-CTQ 2009–09538).

References

[1] Toledano A, Alvarez MI. Lesions and dysfunctions of the nucleus basalis as Alzheimer's disease models: general and critical overview and analysis of the long-term changes in several excitotoxic models. *Curr Alzheimer Res*. 2004; 1:189-214.

[2] World Health Organization. The ICD-10 Classification of Mental and Behavioural Disorders. Clinical descriptions and diagnostic guidelines. Geneva. 1963 (2010 version: http://apps.who.int/classifications /icd10/browse/2010/en)

[3] Minati L, Edginton T, Bruzzone MG, Giaccone G. Current concepts in Alzheimer's disease: a multidisciplinary review. Am *J Alzheimers Dis Other Demen* 2009; 24:95-121.

[4] Toledano A. Hypotheses Concerning the Aetiology of Alzheimer's Disease. *Pharmacopsychiatry* 1988; 21:17-25

[5] Peña F, Gutiérrez-Lerma A, Quiroz-Baez R, Arias C. The Role of beta-Amyloid Protein in Synaptic Function: Implications for Alzheimer's Disease Therapy. *Curr Neuropharmacol* 2006; 4:149-63.

[6] Selkoe DJ. Soluble oligomers of the amyloid beta-protein impair synaptic plasticity and behavior. *Behav Brain Res* 2008; 192:106-13.

[7] Huang HC, Jiang ZF. Accumulated amyloid-beta peptide and hyperphosphorylated tau protein: relationship and links in Alzheimer's disease. *J Alzheimers Dis* 2009; 16:15-27.

[8] Eikelenboom P, Veerhuis R, Scheper W, Rozemuller AJ, van Gool WA, Hoozemans JJ. The significance of neuroinflammation in understanding Alzheimer's disease. *J Neural Transm* 2006; 113:1685-95.

[9] Apelt J, Bigl M, Wunderlich P, Schliebs R. Aging-related increase in oxidative stress correlates with developmental pattern of beta-secretase activity and beta-amyloid plaque formation in transgenic Tg2576 mice with Alzheimer-like pathology. Int *J Develop Neurosci* 2004; 22:475-84.

[10] Toledano A, Álvarez MI. Lesions and dysfunctions of the Nucleus Basalis as Alzheimer's Disease models. General and critical overview and analysis of the long-term changes in several excitotoxic models. *Cur Alzh Res* 2004; 1:189-214.

[11] Opii WO, Joshi G, Head E, Milgram NW, Muggenburg BA, Klein JB, et al. Proteomic identification of brain proteins in the canine model of human aging following a long-term treatment with antioxidants and a program of behavioral enrichment: relevance to Alzheimer's disease. *Neurobiol Aging* 2008; 29: 51-70.

[12] Bernedo V, Insua D, Suárez ML, Santamarina G, Sarasa M, Pesini P. Beta-amyloid cortical deposits are accompanied by the loss of serotonergic neurons in the dog. *J Comp Neurol* 2009; 513: 417-29.

[13] Head E, Moffat K, Das P, Sarsoza F, Poon WW, Landsberg G, et al. Beta-amyloid deposition and tau phosphorylation in clinically characterized aged cats. *Neurobiol Aging* 2005; 26: 749-63.

[14] Cork LC, Powers RE, Selkoe DJ, Davies P, Geyer JJ, Price DL. Neurofibrillary tangles and senile plaques in aged bears. *J Neuropathol Exp Neurol* 1988; 47: 629-41.

[15] Roertgen KE, Parisi JE, Clark HB, Barnes DL, O'Brien TD, Johnson KH. A beta-associated cerebral angiopathy and senile plaques with neurofibrillary tangles and cerebral hemorrhage in an aged wolverine (Gulo gulo). *Neurobiol Aging* 1996; 17: 243-7.

[16] Wallace WC, Bragin V, Robakis NK, Sambamurti K, VanderPutten D, Merril CR, et al. Increased biosynthesis of Alzheimer amyloid precursor protein in the cerebral cortex of rats with lesions of the nucleus basalis of Meynert. *Brain Res Mol Brain Res* 1991; 10: 173-8.

[17] Monzón-Mayor M, Álvarez MI, Arbelo-Galván JF, Romero-Alemán MM, Yanes C, Plaza ML, et al. Long-term evolution of local, proximal and remote astrocyte responses after diverse nucleus basalis lesioning (an experimental Alzheimer model). GFAP immunocytochemical study. *Brain Res*. 2000; 865: 245-58.

[18] Borchelt DR, Ratovitski T, van Lare J, Lee MK, Gonzales V, Jenkins NA, et al. Accelerated amyloid deposition in the brains of transgenic mice coexpressing mutant presenilin 1 and amyloid precursor proteins. *Neuron* 1997; 19: 939-45.

[19] Lüth HJ, Apelt J, Ihunwo AO, Arendt T, Schliebs R. Degeneration of beta-amyloid-associated cholinergic structures in transgenic APP SW mice. *Brain Res* 2003; 977: 16-22.

[20] Rak M, Del Bigio MR, Mai S, Westaway D, Gough K. Dense-core and diffuse Aβ plaques in TgCRND8 mice studied with synchrotron FTIR microspectroscopy. *Biopolymers* 2007; 87: 207-17.

[21] Rodríguez-Casado A, Alvarez I, Toledano A, De Miguel E, Carmona P. Amphetamine effects on brain protein structure and oxidative stress as revealed by FTIR microspectroscopy. *Biopolymers* 2007; 86:4 37-46.

[22] Carmona P, Rodríguez-Casado A, Alvarez MI, De Miguel E, Toledano A. FTIR microspectroscopic analysis of the effects of some drugs on oxidative stress and brain protein structure. *Biopolymers* 2008; 89: 548-54.

[23] Fukui K, Takatsu H, Shinkai T, Suzuki S, Abe K, Urano S. Appearance of amyloid beta-like substances and delayed-type apoptosis in rathippocampus CA1 region through aging and oxidative stress. *J Alzheimers Dis* 2005; 8: 299-309.

[24] Jaya Prasanthi RP, Schommer E, Thomasson S, Thompson A, Feist G, Ghribi O. Regulation of beta-amyloid levels in the brain of cholesterol-fed rabbit, a model system for sporadic Alzheimer's disease. *Mech Ageing Dev* 2008; 129: 649-55.

[25] Woodruff-Pak DS, Agelan A, Del Valle L. A rabbit model of Alzheimer's disease: valid at neuropathological, cognitive, and therapeutic levels. *J Alzheimers Dis* 2007; 11: 371-83.

[26] Montenegro R, Estrada N, Maulini L, Murialdo R. *Biología evolutiva. EA Ediciones*. 2001

[27] Wiley AS. Medical anthropology: a biocultural approach. *University of Southern California*. 2008

[28] Jerison HJ. *The theory of encephalization*. Ann N Y Acad Sci 1977; 299: 146-60.

[29] Sarnat HB, Nestky MG. Evolution of the nervous system. *Oxford Univerity Press London*. 1974

[30] Rilling JK, Insel TR. The primate neocortex in comparative perspective using magnetic resonance imaging. *J Hum Evol* 1999; 37: 191-223.

[31] Schoenemann PT, Sheehan MJ, Glotzer LD. Prefrontal white matter volume is disproportionately larger in humans than in other primates. *Nat Neurosci* 2005; 8: 242-52.

[32] Oró JJ. Evolution of the brain: from behavior to consciousness in 3.4 billion years. *Neurosurgery* 2004; 54: 1287-96.

[33] Barton RA. Evolutionary specialization in mammalian cortical structure. *J Evol Biol* 2007; 20: 1504-11.

[34] Sherwood CC, Subiaul F, Zawidzki TW. A natural history of the human mind: tracing evolutionary changes in brain and cognition. J Anat 2008; 212: 426-54.

[35] Krubitzer L. In search of a unifying theory of complex brain evolution. *Ann N Y Acad Sci* 2009; 1156: 44-67.

[36] Fernández CL, Macias R, Alvarez L. Morpho-functional study brain in aged non-human primates. *Current Primatology* 1993;3:139-44.

[37] Fernández CL, Macias R, Alvarez L, Fermin E. aged nonhuman primates. A multidisciplinary study of brain aging. In: Beregi E, Gergely IA, Rajczi K, editors. Recent advances in Againg Science. Monduzzi Editore; 1993. p. 529-34.

[38] Fernández CL, González O, Antunez I. Cuban preclinical studies in aged monkeys. In: Cacauelos R, Fernandez CL, Giacobini E, Takeda M, editors. Annals of Psychiatry. Basic and Clinical Neurosciences. Barcelona-Philadelphia: *Prous-science*; 1999. p. 1-12.

[39] Voytko, ML. Nonhuman Primates as Models for Aging and Alzheimer's Disease. *Laboratory Animal Science* 1998; 48: 611-7.

[40] Voytko, ML, Tinker, GP. Cognitive function and its neural mechanism in nonhuman primate models of aging, Alzheimer's diseases, and menopause. Frontiers in Bioscience 2004; 9: 1899-914.

[41] Pawlik M, Fuchs E, Walker LC, Levy E. Primate-like amyloid-beta sequence but no cerebral amyloidosis in aged tree shrews. *Neurobiol Aging* 1999; 20: 47-51.

[42] Bons N, Rieger F, Prudhomme D, Fisher A, Krause K-H. Microcebus murinus: a useful primate model for human cerebral aging and Alzheimer's disease? *Genes, Brain and Behavior* 2006; 5: 120-130.

[43] Giannakopoulos P, Silhol S, Jallageas V, Mallet J, Bons N, Bouras C, Delaere P. Quantitative analysis of tau protein-immunoreactive accumulations and beta amyloid protein deposits in the cerebral cortex of the mouse lemur, Microcebus murinus. *Acta Neuropathol* (Berl) 1997; 94: 131-9.

[44] Bachevalier J, Landis LS, Walker LC, Brickson M, Mishkin M, Price DL, Cork LC. Aged monkeys exhibit behavioral deficits indicative of widespread cerebral dysfunction. *Neurobiol Aging* 1991; 12: 99-111.

[45] Uno H, Walker LC. The age of biosenescence and the incidence of cerebral beta-amyloidosis in aged captive rhesus monkeys. *Ann N Y Acad Sci* 1993; 695: 232-5.

[46] Rawlins, R. G. and Kessler, M. J. (eds) The Cayo Santiago macaques. History, behavior and biology. Albany, NY: State University of New York Press, 1986.

[47] McGuire MT. The St Kitts vervet (cercopithecus aethiops). *J Med Primatol* 1974; 3: 285-97.

[48] Palmour RM, Mulligan J, Howbert JJ, Ervin F. Of monkeys and men: vervets and the genetics of human-like behaviors. *Am J Hum Genet* 1997; 61: 481-8.

[49] Lemere CA, Beierschmitt A, Iglesias M, Spooner ET, Bloom JK, Leverone JF, et al. Alzheimer's disease abeta vaccine reduces central nervous system abeta levels in a non-human primate, the Caribbean vervet. *Am J Pathol* 2004; 165:283-97.

[50] Bons N, Mestre N, Ritchie K, Petter A, Podlisny M, Selkoe D. Identification of amyloid beta protein in the brain of the small, short-lived lemurian primate Microcebus murinus. *Neurobiol Aging* 1994; 15: 215-20.

[51] Bartus RT, Dean RL 3rd, Beer B, Lippa AS. The cholinergic hypothesis of geriatric memory dysfunction. *Science* 1982; 217: 408-14.

[52] Terry RD. Alzheimer's disease and the aging brain. *J Geriatr Psychiatry Neurol* 2006; 19: 125-8.

[53] Masliah E, Crews L, Hansen L. Synaptic remodeling during aging and in Alzheimer's disease. *J Alzheimers Dis* 2006; 9: 91-9.

[54] Geula C, Nagykery N, Nicholas A, Wu CK. Cholinergic neuronal and axonal abnormalities are present early in aging and in Alzheimer disease. *J Neuropathol Exp Neurol* 2008; 67: 309-18.

[55] Brizee, KR, Ordy JM, Bartus RT. Localization of cellular changes within multimodal sensory regions in aged monkey brain: possible implications for aged-related cognitive less. *Neurobiol Aging* 1980; 1: 45-52.

[56] Peters A, Leahu D, Moss MB, McNally KJ. The effects of aging on area 46 of the frontal cortex of the rhesus monkey. *Cereb Cortex* 1994; 4: 621-35.

[57] Peters A. Structural changes in the normally aging cerebral cortex of primates. *Prog Brain Res* 2002; 136: 455-65.

[58] Vicent SL, Peters A, Tiggers J. Effects of aging on the neurons within area 17 of rhesus monkey cerebral cortex. Anat Rec. 1989, 223: 329-341.

[59] Peters A, Moss MB, Sethares C. The effects of aging on layer 1 of primary visual cortex in the rhesus monkey. *Cereb Cortex*. 2001; 11: 93-103.

[60] Keuker JI, Luiten PG, Fuchs E. Preservation of hippocampal neuron numbers in aged rhesus monkeys, *Neurobiol Aging* 2003; 24: 157–165

[61] Cruz L, Roe DL, Urbanc B, Inglis A, Stanley HE, Rosene DL. Age-related reduction in microcolumnar structure correlates with cognitive decline in ventral but not dorsal area 46 of the rhesus monkey. *Neuroscience* 2009; 158: 1509-20.

[62] Page TL, Einstein M, Duan H, He Y, Flores T, Rolshud D, et al. Morphological alterations in neurons forming corticocortical projections in the neocortex of aged Patas monkeys. *Neurosci Lett*, 2002; 317: 37-41.

[63] Uemura, E. Aged-related changes in the prefrontal cortex of Maccaca mulatta: synaptic density. *Exp Neurol* 1980; 69:164-72

[64] Cupp CJ, Uemura E. Aged-related changes in prefrontal cortex of Maccaca mulatta: quantitative analysis of dendritic branching patterns. *Exp Neurol* 1980; 69: 143-63.

[65] Bertoni-Freddari C, Fattoretti P, Giorgetti B, Grossi Y, Balietti M, et al. Synaptic pathology in the brain cortex of old monkeys as an early alteration in senile plaque formation. *Rejuvenation Res* 2006; 9: 85-8.

[66] Peters A, Sethares C, Luebke JI. Synapses are lost during aging in the primate prefrontal cortex. *Neuroscience* 2008, 152: 970-81.

[67] Dumitriu D, Hao J, Hara Y, Kaufmann J, Janssen WG, Lou W, et al. Selective changes in thin spine density and morphology in monkey prefrontal cortex correlate with aging-related cognitive impairment. *J Neurosci* 2010; 30: 7507-15

[68] Voytko ML, Sukhov RR, Walker LC, Breckler SJ, Price DL, Koliatsoso VE. Neuronal number and size are preserved in the nucleus basalis of aged rhesus monkeys. *Dementia* 1995; 6: 131-41.

[69] Stroessner-Johnson HM, Rapp PR, Amaral DG. Cholinergic cell loss and hypertrophy in the medial septal nucleus of the behaviorally characterized aged rhesus monkey. *J Neuroscience* 1992; 12: 1936-44.

[70] Kemper TL. The relationship of cerebral cortical changes to nuclei in the brainstem. *Neurobiol Aging* 1993; 14: 659-60.

[71] Rosene DL. Comparing aged-related changes in the basal forebrain and hippocampus of the rhesus monkey. *Neurobiol Aging* 1993; 14: 669-70.

[72] Sandell JH, Peters A. Disrupted myelin and axon loss in the anterior commissure of the aged rhesus monkey. *J Comp Neurol* 2003; 466: 14-30.

[73] Price DL, Koo EH, Sisodia SS, Martin LJ, Walker LC, Kitt CA, et al. Brain abnormalities in aged monkeys: a model sharing features with AD. In: Maurer K, Riederer P, Beckam H, editors. *Alzheymer's diseasse. Springer Verlag*; 1990. p. 141-50

[74] Price DL, Martin LJ, Sisodia SS, Wagster MV, Koo EH, Walker LC, et al. Aged non-human primates: an animal model of age-associated neurodegenerative disease. *Brain Pathol* 1991; 1: 287-96.

[75] Silhol S, Calenda A, Jallageas V, Mestre-Frances N, Bellis M, Bons N. Beta-Amyloid protein precursor in Microcebus murinus: genotyping and brain localization. *Neurobiol Dis* 1996; 3: 169-82.

[76] Uno, H. Aged-related pathology and biosenescent markers in captive rhesus macaques. *Age* 1997; 20: 1-13.

[77] Gearing M, Rebeck GW, Hyman BT, Tigges J, Mirra SS. Neuropathology and apolipoprotein E profile of aging chimpanzees: implications for Alzheimer disease.*Proc Natl Acad Sci U S A* 1994; 91: 9382-6.

[78] Poduri A, Gearing M, Rebeck GW, Mirra SS, Tigges J, Hyman BT. Apolipoprotein E4 and beta amyloid in senile plaques and cerebral blood vessels of aged rhesus monkeys. *Am J Pathol* 1994; 144: 1183-7.

[79] Calenda A, Mestre-Frances N, Czech C, Pradier L, Petter A, Perret M, et al Bons N, Bellis M. Cloning of the presenilin 2 cDNA and its distribution in brain of the primate, Microcebus murinus: coexpression with betaAPP and Tau proteins. *Neurobiol Dis* 1998; 5:323-33.

[80] Gearing M, Tigges J, Mori H, Mirra SS. beta-Amyloid (A beta) deposition in the brains of aged orangutans. *Neurobiol Aging*. 1997; 18: 139-46.

[81] Lemere CA, Oh J, Stanish HA, Peng Y, Pepivani I, Fagan AM, et al. Cerebral amyloid-beta protein accumulation with aging in cotton-top tamarins: a model of early Alzheimer's disease? *Rejuvenation Res*. 2008; 11: 321-32.

[82] Maclean CJ, Baker HF, Ridley RM, Mori H. Naturally occurring and experimentally induced beta-amyloid deposits in the brains of marmosets (Callithrix jacchus). *J Neural Transm*. 2000; 107: 799-814.

[83] Ridley RM, Baker HF, Windle CP, Cummings RM. Very long term studies of the seeding of beta-amyloidosis in primates. *J Neural Transm*. 2006; 113: 1243-51.

[84] Walker LC. Animal models of cerebral beta-amyloid angiopathy. *Brain Res Brain Res Rev* 1997; 25: 70-84.

[85] Cai Y, Xiong K, Zhang XM, Cai H, Luo XG, Feng JC, et al. Beta-Secretase-1 elevation in aged monkey and Alzheimer's disease human cerebral cortex occurs around the vasculature in partnership with multisystem axon terminal pathogenesis and beta-amyloid accumulation. *Eur J Neurosci* 2010; 18: 1223-38

[86] Rosen RF, Farberg AS, Gearing M, Dooyema J, Long PM, Anderson DC, et al. Tauopathy with paired helical filaments in an aged chimpanzee. *J Comp Neurol.* 2008; 509: 259-70.

[87] Oikawa N, Kimura N, Yanagisawa K. Alzheimer-type tau pathology in advanced aged nonhuman primate brains harboring substantial amyloid deposition. *Brain Res.* 2010; 1315: 137-49.

[88] Kimura N, Tanemura K, Nakamura S, Takashima A, Ono F, Sakakibara I, et al. Age-related changes of Alzheimer's disease-associated proteins in cynomolgus monkey brains. *Biochem Biophys Res Commun* 2003; 310: 303-11.

[89] Kimura N, Nakamura S, Goto N, Narushima E, Hara I, Shichiri S, Saitou K, Nose M, Hayasi T, Kawamura S, Yoshikawa Y. Senile plaques in an aged western lowland gorilla. *Exp Anim.* 2001; 50(1):77-81.

[90] Shah P, Lal N, Leung E, Traul DE, Gonzalo-Ruiz A, Geula C. Neuronal and axonal loss are selectively linked to fibrillar amyloid-{beta} within plaques of the aged primate cerebral cortex. *Am J Pathol.* 2010; 177: 325-33.

[91] Price DL, Cork LC, Struble RG, Whitehouse PJ, Kitt CA, Walker LC. The funtional organization of the basal forebrain cholinergic system in primates and the role of its system in Alzheymer´s diseasse. *Ann NY Acad Sci.* 1985; 444:287-95.

[92] Price DL, Kitt CA, Struble RG, Whitehouse PJ, Cork LC. Neurobiological studies of transmitter systems in aging and Alzheymer type dementia. *Ann NY Acad Sci* 1985; 444:35-51.

[93] Bertoni-Freddari C, Fattoretti P, Giorgetti B, Grossi Y, Balietti M, Casoli T, et al. Alterations of synaptic turnover rate in aging may trigger senile plaque formation and neurodegeneration. *Ann NY Acad Sci* 2007; 1096: 128-37.

[94] González-Fraguela ME, Fernández CL, Robinson MA, Delgado R. A study of the oxidative stress indicators in elderly monkeys. *Rev. Neurol.* 2000 31:1012-5.

[95] Trouche SG, Maurice T, Rouland S, Verdier JM, Mestre-Francés N. The three-panel runway maze adapted to Microcebus murinus reveals age-related differences in memory and perseverance performances. *Neurobiol Learn Mem* 2010; 94: 100-6

[96] Trouche SG, Asuni A, Rouland S, Wisniewski T, Frangione B, Verdier JM, et al. Antibody response and plasma Abeta1-40 levels in young Microcebus murinus primates immunized with Abeta1-42 and its derivatives. *Vaccine* 2009; 27: 957-64.

[97] Kanemaru K, Iwatsubo T, Ihara Y. Comparable amyloid beta-protein (A beta) 42(43) and A beta 40 deposition in the aged monkey brain. *Neurosci Lett* 1996; 214: 196-8.

[98] Choudhury, A.U. Status and conservation of the stump-tailed macaque Macaca arctoides in India. *Primate. Report*, 2002 63:63-72.

[99] Kessler MJ. Establishment of the Cayo Santiago colony. *P R Health Sci J.* 1989 8:15-7.

[100] Morales L, Baez MM, Fernández CI, López V, Díaz L. Seguimiento evolutivo con EEG cuantitativo de la infusion ICV de factor de crecimiento nervioso en monos envejecidos. *Rev. Neurol* 1997; 25(142):850-5.

[101] Baez MM, Morales L, Fernández CI, Cabrera I, Padilla E, González O, et al. The aged monkey: Electrophysiological follow-up of Nerve Growth Factor effects. In: Cacabelos R, Fernández CI, Giacobini E, Takeda M, editors. Annals of Psychiatry. Basic and Clinical Neurosciences. Barcelona-Philadelphia: Prous-science; 1999. p. 309-11.

[102] Buccafusco JJ, Terry AV, Murdoch PB. A computer-assisted cognitive test battery for aged monkeys. *J Mol Neurosci* 2002; 19:179-85.

[103] Squire LR, Zola-Morgan S, Chen KS. Human amnesia and animal models of amnesia: performance of amnesic patients on test designed for the monkey. Behav. *Neurosci.* 1988; 102:210-217.

[104] Perryman KM, Fitten LJ. Delayed Matching to sample performance during a doubl blind trial of tacrine (THA) and lecithin in patients with Alzheimer's diseasse. *Life Sci* 1993; 53:479-86.

[105] Rapp, Amaral DG. Indivudual differences in the cognitive and neurobiological consequences of normal aging. *Trends Neurosci* 1992; 15:340-5.

[106] Dwight P, Siwak CT, Estrada J, Holowachuck D, milgram NW. Effects of Age on measures of complex working memory span in the beagle dog (Canis familiaris) using two versions of a spatial list learning paradigm. *Learn Mem* 2003; 10:148-60

[107] Glisky, El. Changes in Cognitive Function in Human Aging. In: Riddle DR, Editor. Brain Aging: Models, Methods, aand Mechanisms. Boca Raton (FL): *CRC Press*; 2007. Chapter 1.

[108] Mark B. Moss, Tara L: Moore, Steophen P. Schettler, Ronald Killiany, And Douglas Rosene. Susccessful vs. Unsuccessful Aging in the Rhesus Monkey. In: Riddle DR, editor. Brain Aging: Models, Methods, aand Mechanisms. Boca Raton (FL): *CRC Press*; 2007. Chapter 2. Frontiers in Neuroscience.

[109] Carmona P, Molina M, Calero M, Bermejo-Pareja F, Martínez-Martín P, Álvarez MI, Toledano A. Infrared analysis of mononuclear leukocytes in peripheral blood from Alzheimer's disease patients. *Anal Bioanal Chem.* 2012 (in press). DOI 10.1007/s00216-011-5669-9

[110] Voytko ML. Impairements in acquisition and reversal of two-choice discriminations by aged Rhesus monkeys. *Neurobiol Aging* 1999; 20: 617-27.

[111] Presty SK, Bachevalier J, Walker LC, Price DL, Cork LC. Age Differences in recognition memory of the rhesus monkey (Macaca mulatta). *Neurobiol Aging* 1987; 8:435-40.

[112] Bartus RT, Fleming D, Johnson HR. Aging in the rhesus monkey: debiliting effects on short-term memory. *J Gerontol* 1978; 33: 858-71.

[113] Mariott JG, Abelson JS. Age diferences in short-term memory of test-sophisticated rhesus monkeys. *Age* 1980; 3: 7-9.

[114] Rapp P.R. and D.G. Amaral: Evidence for task-dependent memory dysfunction in the aged monkey. *J Neurosci* 1989; 9: 3568-76.

[115] Flicker C, Bartus RT, Crook TH, Ferris SH. Effects of aging and dementia upon recent visuospatial memory. *Neurobiol Aging* 1984; 5: 275-283.

[116] Davis RT, Bennett CA, Weisenburger WP. Repeated measurements of forgetting by rhesus monkeys (Macaca mulatta). *Percept Mot Skills* 1982; 55: 703-9.

[117] Johnstone EM, Chaney MO, Norris FH, Pascual R, Little SP. Conservation of the sequence of the Alzheimer's disease amyloid peptide in dog, polar bear and five other mammals by cross-species polymerase chain reaction analysis. *Brain Res Mol Brain Res* 1991; 10: 299-305.

[118] Duyckaerts C, Delatour B, Potier MC. Classification and basic pathology of Alzheimer disease. *Acta Neuropathol* 2009; 118:5-36.

[119] Walker LC, Rosen RF, Levine H 3rd. Diversity of Abeta deposits in the aged brain: a window on molecular heterogeneity? *Rom J Morphol Embryol 2008*; 49: 5-11.

[120] Attems J, Jellinger K, Thal DR, Van Nostrand W. Review: sporadic cerebral amyloid angiopathy. *Neuropathol Appl Neurobiol.* 2011;37:75-93.

[121] Chambers JK, Kuribayashi H, Ikeda S, Une Y. Distribution of neprilysin and deposit patterns of Abeta subtypes in the brains of aged squirrel monkeys (Saimiri sciureus). *Amyloid* 2010; 17: 75-82.

[122] Holzer M, Craxton M, Jakes R, Arendt T, Goedert M. Tau gene (MAPT) sequence variation among primates. *Gene.* 2004 341:313-22.

[123] Kraska A, Dorieux O, Picq JL, Petit F, Bourrin E, Chenu E, et al. Age-associated cerebral atrophy in mouse lemur primates. *Neurobiol Aging* 2011; 32: 894-906.

[124] Ichinohe N, Hayashi M, Wakabayashi K, Rockland KS. Distribution and progression of amyloid-beta deposits in the amygdala of the aged macaque monkey, and parallels with zinc distribution. *Neuroscience* 2009; 159: 1374-83.

[125] Craik FIM, Byrd M, Swanson JM. Patterns of memory loss in three elderly samples. *Psychol Aging* 1987; 2: 79-86.

[126] Zelinski EM, Gilewski MJ, Scahie KW. Individual differences in cross-sectional and 3-year longitudinal memory performance across the adult life span. *Psychol Aging* 1993; 8: 176-86.

[127] Morse CK. Does variability increase with age? An archival study of cognitive measures. *Psychol Aging* 1993; 8: 156-64.

[128] Moss MB, Moore TL, Schettler SP, Killiany R, Rosene D. Successful vs. unsuccessful Aging in the Rhesus Monkey. En: Riddle DR, editor. Brain Aging: Models, Methods, and Mechanisms. *Frontiers in Neuroscience* Cap 2. Boca Raton (FL): CRC Press, 2007

[129] Rosen RF, Walker LC, Levine H 3rd. PIB binding in aged primate brain:Enrichment of high-affinity sites in humans with Alzheimer's disease. *Neurobiol Aging* 2009; 32: 223-34

[130] Rapoport SI. Hypothesis: Alzheimer's disease is a phylogenetic disease. *Med Hypotheses* 1989; 29: 147-50.

[131] Vamathevan JJ, Hasan S, Emes RD, Amrine-Madsen H, Rajagopalan D, Topp SD, et al. The role of positive selection in determining the molecular cause of species differences in disease. *BMC Evol Biol* 2008; 8: 273.

[132] Albert MS. Neuropsychological and neurophysiological changes in healthy adult humans across the age range. *Neurobiol Aging* 1993; 14:623.

[133] Perls TT. The different paths to 100. Am J Clin Nutr 2006; 83: 484S.

[134] Forlenza OV, Diniz BS, Gattaz WF. Diagnosis and biomarkers of predementia in Alzheimer's disease. *BMC Med* 2010; 8: 89.
[135] Nelson PT, Stefansson K, Gulcher J, Saper CB. Molecular evolution of tau protein: implications for Alzheimer's disease. *J Neurochem* 1996; 67: 1622-32.
[136] Bufill E, Blesa R. Enfermedad de Alzheimer y evolución cerebral. ¿Es la enfermedad de Alzheimer un ejemplo de pleiotropía antagonística?. *Rev Neurol* 2006; 42: 25-33.
[137] Finnerty JR, Wang WX, Hébert SS, Wilfred BR, Mao G, Nelson PT. The miR-15/107 group of microRNA genes: evolutionary biology, cellular functions, and roles in human diseases. *J Mol Biol* 2010; 402: 491-509.
[138] Wallace DC Mitochondrial DNA sequence variation in human evolution and disease Proc Natl Acad Sci U S A 1994; 91: 8739-46.

In: Primates
Editors: E. F. Hughes and M. E. Hill
ISBN: 978-1-62081-498-7

Chapter III

The Rotator Cuff Muscles in Hominoidea: Evolution and Adaptations to Different Types of Locomotion

***G. Bello-Hellegouarch*[1], *J. M. Potau*[2], *J. Arias-Martorell*[1], *J. F. Pastor*[3], *R. Diogo*[4] and *A. Pérez-Pérez*[1]**

[1]Barcelona University, Anthropology Unit of Animal Biology Department, Barcelona, Spain

[2]Barcelona University, Unit of Human Anatomy and Embryology, Barcelona, Spain

[3]Valladolid University, Department of anatomy, Leon, Spain

[4]Howard University, Department of anatomy, Washington, DC, US

Abstract

The rotator cuff muscles (subscapularis, supraspinatus, infraspinatus and teres minor) are the principal stabilizers of the glenohumeral joint in primates. This function is particularly important in hominoids due to their orthograde corporal pattern, characterized by a dorsal position of the scapulae, a high mobility of the glenohumeral joint and the presence of tensile forces that affect the shoulder region during locomotion. These functional and physiological conditions imply an special morphology of the rotator cuff muscles in hominoids, which present an increased global mass and important differences in the proportion of the different muscles with respect to the more quadrupedal pronograde primates. Although sharing some similarities, the different general that form the Hominoidea have developed different types of locomotion such as brachiation (*Hylobates*), arm-swinging and vertical climbing (*Pongo*), knuckle-walking (*Gorilla* and *Pan*) and bipedalism (*Homo*). Thus, the morphology of the rotator cuff muscles in these genera is also related to the functional requirements of these types of locomotion. In this chapter we will discuss the anatomical and functional aspects of the rotator cuff muscles in hominoids by focusing on the evolutionary transformations and adaptations related to the different locomotor modes developed by hylobatids, orangutans, gorillas, chimpanzees and modern humans.

Introduction

As stressed by Gebo (2010: p. 530), "life in the canopy is a constant stream of body adjustments". Environmental changes result in the alteration of the requirements for posture and locomotion, requirements that are reflected in anatomical adaptations (Sonnabend and Young, 2009), closely related to the type of locomotion used to move through the canopy.

It has been proposed that hominoids became different from other primates (and from mammals in general) due to their specific adaptations to an arboreal environment where the capacity to a below-branch forelimb-dominated arboreal locomotion was an advantage in survival (Le Gros Clark, 1959; Larson, 1993; Fleagle, 1999; Sonnabend and Young, 2009). Thus, extant apes are characterized by some locomotor specializations, sharing body morphologies associated to their locomotor behaviors (Ward, 2007), especially those related to suspension and orthograde climbing, many of them still seen in modern humans. Crucial anatomical features characterizing the Hominoidea are seen in the thoracic region and principally the shoulder, which displays a significant amount of morphological variation related to an accommodation to different habits and habitats (Figure 1; Oxnard, 1967; Larson, 1993). Pronograde primates (and mammals in general) share a pronograde back, a deep dorso-ventrally thorax, a scapula located in the parasagittal plane lateral to the rib cage and a glenohumeral joint adapted for quadrupedal locomotion (both arboreal and terrestrial) with movements restricted to the parasagittal plane (Schultz, 1930; Roberts, 1974; Aiello and Dean, 1990; Larson, 1993;

Sonnabend and Young, 2009; Gebo, 2010). In contrast, living ape and human backs are adapted to vertical and erect postures, sharing an erect back (what is commonly known as orthogrady) and a thorax that is shallow antero-posteriorly and wide laterally. Their scapula is displaced to a dorsal position rather than along the side of the ribcage as in quadrupedal primates, forcing the shoulder to the sides and away from the midline of the body, developing a more mobile and less stable glenohumeral joint with the glenoid fossa facing laterally rather than ventrally, which can additionally rotate and move in the coronal plane, allowing more freedom of mobility with multi-directional movements (the shoulder joint is the most freely movable in the body of hominoids; Figure 1; Keith, 1923; Schultz, 1930; Roberts, 1974; Aiello and Dean, 1990; Larson, 1993; Fleagle, 1999; Sonnabend and Young, 2009; Gebo, 2010). The shoulder acquired the required mobility to reach the irregular supports of a discontinuous arboreal substrate (Larson, 1993), allowing more specialized forms of locomotion, including brachiation, vertical climbing, suspensory posture and knuckle-walking, and, in modern humans, the use of the hand for complex manipulation (Schultz, 1961; Corruccini and Ciochon, 1976; Aiello and Dean, 1990; Larson, 1993; Sonnabend and Young, 2009). However, it should be noted that the increase of mobility reduces the stability of the glenohumeral joint (Roberts, 1974; Larson, 1993); this means that the shoulder region will reflect a compromise between the free mobility demands and the need for stability, facing the disruptive forces generated during locomotion and the support of the body weight (Larson, 1993).

Within extant apes, hylobatids are the more arboreal ones, being true brachiators due to their particular type of locomotion where the arms are usually fully extended above their head in order to suspend and propel the body through the trees (Napier, 1963; Tuttle, 1975; Fleagle, 1999). Gorillas, chimpanzees and orangutans are considered modified brachiators (Aiello and Dean, 1990) because they use their lower limbs too to a greater or lesser extent in order to provide some support for the body. Orangutans, as hylobatids, almost never move on the ground, but they differ in that the former show a larger body size, moving cautiously through the upper levels of the canopy with quadrumanous climbing and arm-hanging (Fleagle, 1999; Thorpe and Crompton, 2005, 2006; Ward, 2007). Common chimpanzees are arboreal especially when they forage, hunt and sleep, but, in contrast to orangutans, they spend much of their time as terrestrial knuckle-walkers (Goodall, 1965; Fleagle, 1999). The locomotor behavior of bonobos is less well studied, but it seems to be somewhat similar to that of common chimpanzees (Doran, 1993), although it is thought (Susman, 1984; Doran and Hunt, 1994) that bonobos often engage in more arboreal leaping, diving and arm-swinging. Gorillas are the biggest modern primates and are almost exclusively terrestrial, although sporadically in certain habitats they climb trees, frequently for feeding and sleeping (Tuttle and Watts, 1985; Remis, 1995; Fleagle, 1999). Knuckle-walking, the primary type of locomotion of the African apes, is a quadrupedalism that has been secondarily imposed on a forelimb that was being primarily adapted to suspensory postures and locomotion (Larson and Stern, 1987). Thereby, gorillas and chimpanzees may have evolved to acquire those functions related to quadrupedal weight bearing, but without losing the previous arm-swinging and suspensory functions (Oxnard, 1969). Modern humans, with their bipedal locomotion, lost any adaptations needed solely for locomotion, accentuating those ones that permitted them to transform the upper limb into an almost exclusively manipulative extremity possessing important tactile, prehensile, manipulatory, communicative and agonistic functions (Ziegler, 1964; Oxnard, 1969; Fleagle, 1999).

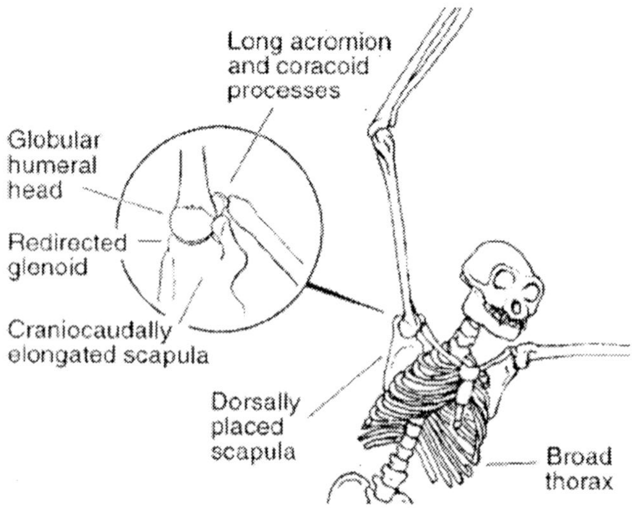

Figure 1. Characteristic skeletal features of the shoulder region of extant apes (modified from Fleagle, 1999).

The comparative morphology of the primate upper limb, and more specifically, of the scapula, has been very useful to establish functional differences in behaviour among hominoids (Ashton and Oxnard, 1964; Roberts, 1974; Corruccini and Ciochon, 1976). This focus on the primate scapula is due to the fact that this is the most variable bone of the forelimb (Oxnard, 1967), and, as it is suspended almost completely by muscles it is supposed to reflect, at least in some of its variations, the deep relationship with the function of the upper limb (Figure 2; Inman *et al*, 1944; Oxnard, 1967) and the forces this limb has to endure during locomotor behaviors. For example, the shoulder of brachiators is mainly subject to tensile forces, while the shoulder of quadrupeds is mainly subject to compressive forces (Oxnard, 1967, 1969; Roberts, 1974; Larson, 1993).

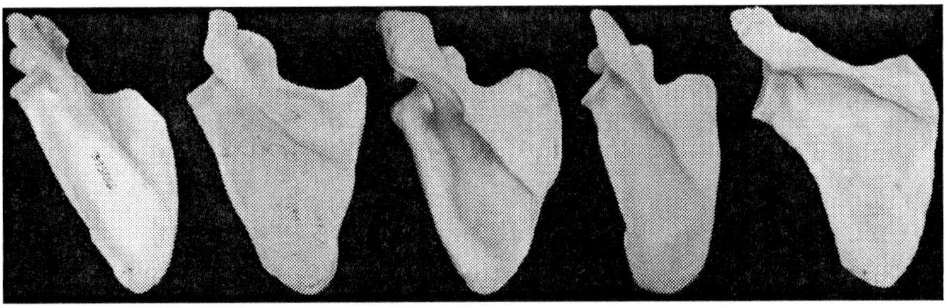

Figure 2. Scapulae of the different groups of the Hominoidea. From left to right: *Hylobates* (ref.395156, from the Royal Belgian Institute of Natural Sciences), *Pongo* (ref.94-0703 from the Natural History Museum of Barcelona), *Gorilla* (ref.2003-0544 from the Natural History Museum of Barcelona), *Pan* (ref.95.0258 from the Natural History Museum of Barcelona) and *Homo* (ref.038 from the Unit of Human Anatomy and Embryology of the University of Barcelona).

Before presenting our own data and describing in detail the differences between the shoulder, and more specifically the scapula and scapulohumeral muscles of the different hominoid taxa, it is convenient to carry out a brief, simplified introduction to some of the muscles attached onto the scapula.

Brief Introduction to the Rotator Cuff Muscles

As we explained above, the emphasis on the mobility of the hominoid shoulder joint results in a reduction of the stability of the joint (Roberts, 1974; Larson, 1993), thus requiring additional stabilization to prevent its dislocation. This is easy to understand when we observe the bony architecture of the glenohumeral joint, with its large articulating humeral head and its relatively small glenoid surface (Terry *et al*, 2000). This stabilization is especially important during behaviors that imply large tensile stress in the joint, so it is not maintained solely by static osseoligamentous structures but also, and principally, by dynamic stabilizers (Larson and Stern, 1986; Terry *et al*, 2000; Jenkins, 2009). The dynamic stabilization is accomplished by a group of muscles known as the rotator cuff or the musculotendinous cuff of the shoulder (Hess, 2000; Jenkins, 2009), whose tendons are closely related to the capsule of the joint anteriorly, above and posteriorly (Terry *et al*, 2000; Jenkins, 2009). Anteriorly, the cuff is made up of the subscapularis muscle (Figure 3), while posteriorly, from a superior to an inferior position, it is made up of the supraspinatus, infraspinatus and teres minor muscles (Figure 4).

The supraspinatus muscle originates from the dorsal surface of the scapula above the spine, known as supraspinous fossa, and from the fascia covering the muscle itself (Figure 4). It passes first over the top of the glenohumeral joint, inserting on the proximal part of the great tubercle (Ashton and Oxnard, 1963; Jenkins, 2009). Before its insertion, the fibers converge upon a short and stout tendon that adheres to the glenohumeral capsule. Between the supraspinatus and the acromion we find the subacromial bursa. The supraspinatus is principally an abductor of the arm, assisting the deltoid in this movement (Larson and Stern, 1986; Aiello and Dean, 1990; Jenkins, 2009), and it is also important to resist humeral displacement (Larson and Stern, 1986, 1987).

The infraspinatus muscle originates from the infraspinous fossa and from its covering fascia, and inserts on the dorsal facet of the great tubercle, inferiorly to the the supraspinatus insertion (Figure 4). The teres minor originates from about the upper two thirds of the posterior surface of the lateral border of the scapula, and from a septa between it and both the infraspinatus above and the teres major below (Figure 4). This muscle inserts distally on the greater tubercle, directly inferior to the insertion of the infraspinatus, after adhering to the dorsal aspect of the glenohumeral capsule. Both the teres minor and the infraspinatus are primarily lateral rotators of the humerus, and they also participate in the maintenance of the position of the humeral head during other movements of the arm (Ashton and Oxnard, 1963; Larson and Stern, 1986; Aiello and Dean, 1990; Jenkins, 2009).

The subscapularis muscle originates from most of the anterior surface of the scapula, known as subscapular fossa, and passes across the front of the glenohumeral joint to its insertion on the lesser tubercle of the humerus (Figure 3). Before inserting, the fibers converge upon a short and thick tendon which adheres to the anterior part of the glenohumeral capsule (Ashton and Oxnard, 1963). Between this muscle and the neck of the scapula lies the subscapularis bursa, usually opening into the synovial cavity of the glenohumeral joint. The subscapularis muscle produces mainly medial rotation, as well as protection to prevent anterior dislocation of the head of the humerus (Aiello and Dean, 1990; Jenkins, 2009).

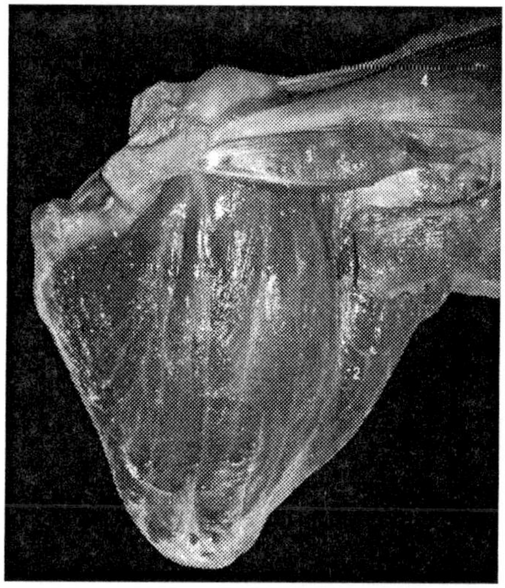

Figure 3. Anterior view of the dissection of *Pan troglodytes* showing the anterior side of the rotator cuff muscles and some other muscles of the shoulder and arm. 1: Subscapularis. 2: Teres major. 3: Coracobrachialis. 4: Biceps brachii (caput breve).

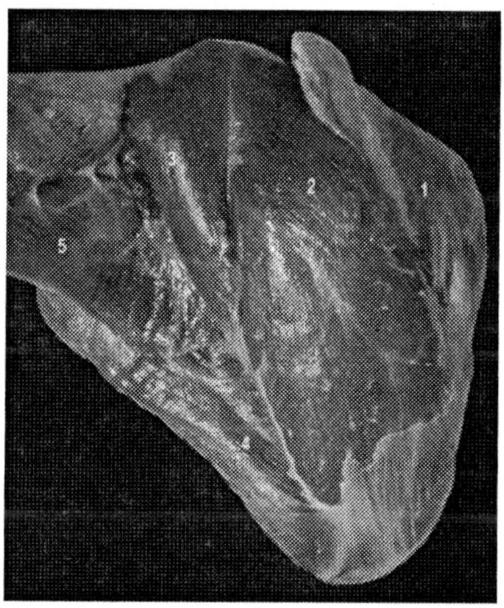

Figure 4. Dorsal view of the dissection of a male *Pan troglodytes* showing the dorsal side of the rotator cuff muscles and some other muscles of the shoulder and arm. 1: Supraspinatus. 2: Infraspinatus. 3: Teres minor. 4: Teres major. 5: Triceps brachii (caput longum).

Thus, in general, the rotator cuff contributes, on the one hand, to the mobility of the shoulder, and on the other hand, to the dynamic stabilization of the shoulder (Roberts, 1974; Sonnabend and Young, 2009), by holding the head of the humerus in the glenoid cavity with a dynamic steering mechanism (Terry *et al*, 2000; Jenkins, 2009).

Accordingly, the architectural arrangement of the rotator cuff muscles fibers is consistent with this stabilizing role, as they seem to be configurated for force production rather than excursion (Ward *et al*, 2006).

Since this increasing need for stability seems to be related to the increased range of motion of the glenohumeral joint, i.e. to the increased use of the arm away from the sagittal plane (Sonnabend and Young, 2009), it seems logical to think that the appearance of a true rotator cuff, with tendons closely related to the capsule and to each other, should be associated with the ability to carry out regular overhead activity.

Indeed, some authors (e.g. Sonnabend and Young, 2009) consider that only primates with an important amount of arboreal locomotion develop true rotator cuff, such as hominoids and some arboreal New World monkeys.

As we will be seen throughout this chapter, the evolution of the bony skeleton of the shoulder girdle has been well described, specially the morphology of the scapula, while there is little information related to the soft tissue, and in particular to the rotator cuff (Sonnabend and Young, 2009).

Many of the measures and studies related with the scapula are not focused on the rotator cuff mechanism, that is, many authors employ some anatomical measures only in order to discriminate between, for example, locomotor groups, without focusing on where the differences are, or which is the reason (related to the musculature) for those differences.

However, as we shall see later, in the last years the musculature *per se* is starting to be the subject of more studies, including gross anatomical, electromyographic and myosine analyses.

Below we will thus provide a detailed review of what is currently known about the rotator cuff muscles in hominoids, based on our own studies and on those of other authors, including classical methods such as the analysis of linear measures of the scapula to the most innovative techniques such as 3D geometric morphometrics and the analysis of the expression of the myosin heavy chain (MHC) isoform.

State of the Art Concerning the Rotator Cuff Muscles

After several empirical studies, from the removal of muscles in rats (e.g. Wolffson, 1950) to genetically modified mice (e.g. Green *et al*, 2011), it has been clearly demonstrated that the rotator cuff muscles have a significant influence on the scapular shape (Howell, 1917; Wolffson, 1950; Roberts, 1974; Green *et al*, 2011), enlarging or reducing their relative sizes and generating stresses that have a great impact in the scapular structure (Inman *et al*, 1944; Roberts, 1974). Thus, the adaptations of size, shape and position of the muscles according to the functional demands of the upper limb can cause adaptations in the position and orientation of the insertions on the bones and joints over which the muscles act, increasing the mechanical efficiency of arm-raising, compared to quadrupeds (Inman *et al*, 1944; Roberts, 1974; Ashton and Oxnard, 1964; Oxnard, 1967). This relationship between the scapular shape and the muscular features of the rotator cuff justifies that the scapula is usually studied in order to obtain muscular information from the osteological data, an easier way to research, due to the limitations (or impossibility in the case of extinct species) related to the direct study of the muscles (principally, the difficulty in obtaining cadavers of primates).

In relation to the rotator cuff mechanism, the model proposed by Inman *et al* (1944) was assumed to be correct and most of the scapular shape variables employed in several studies were derived from this model (see, e.g. Ashton & Oxnard, 1963). According to this model in modern humans there is a force couple operating at the glenohumeral joint, the upper unit of the model being composed by the deltoid and supraspinatus, the lower unit including the infraspinatus, teres minor and subscapularis; the pressure and friction of the humeral head at the glenoid was also taken into account. According to Inman, the deltoid, assisted by the supraspinatus, provides the power to elevate the arm, while the lower unit acts collectively to hold the humeral head down thus preventing the tendency for the deltoid to produce superior humeral displacement. This model was considered correct for many years, not only for modern humans, but for the great apes (Tuttle and Basmajian, 1978a, 1978b), implying that the glenohumeral joint integrity was maintained solely by osseoligamentous structures (Tuttle and Basmajian, 1978a, 1978b). However, further EMG analyses of non-human primate shoulder muscle function (Larson and Stern 1986, 1987) have shown that this model of force couple is not correct in at least chimpanzees, and probably also in other hominoids, beause these analyses emphasized the distinct role of each member of the rotator cuff in free arm movements (Larson and Stern, 1986). Furthermore, the rotator cuff has a second and equally important function in stabilizing the shoulder joint, a role proposed by Roberts in 1974. Therefore, considering that each muscle plays a distinct role in the rotator cuff mechanism for the free movement of the arm, we will try to make its understanding clear and simple, by examining separately each of the rotator cuff muscles.

Supraspinatus

The supraspinatus is the better studied of the rotator cuff muscles, probably due to its importance in stabilizing the joint, in abducting the arm, and also because of its clinical implications (e.g. the subacromial impingement, a typical human disease, see Potau *et al*, 2007). As explained above, most of the studies of the rotator cuff muscles are based on the analysis of the shape and size of their origins, especially in the case of the supraspinatus and the infraspinatus muscles. Thus, much of the information available concerns the supraspinous and infraspinous fossae, obtained in different ways, e.g. multivariate analyses of lineal measures (Ashton and Oxnard, 1964; Ashton, Oxnard and Spence, 1965; Oxnard, 1967; Oxnard and Neely, 1969; Roberts, 1974) and geometric morphometric analyses of the scapula (Young, 2002, 2008). This is a useful way to obtain information about a muscle when this muscle is absent (e.g. fossils; osteological collections), but one must be cautious because our own studies indicate that the size of the muscle and of its origin (in this case, the fossae of the scapula) may not be as closely related as previously though (Bello-Hellegouarch *et al*, unpublished data).

It is clear that the morphology of the supraspinatus varies with the type of locomotion. Hominoids are associated with relatively well-developed supraspinous fossa (Frey, 1923; Schultz, 1930; Inman *et al*, 1944; Roberts, 1974; Corruccini and Ciochon, 1976; Shea, 1986; Larson, 1993), and, by implication, with well developed supraspinatus muscles. This seems to be associated to the presence of relatively large, heavy and powerful limbs with a great need for glenohumeral joint stabilization during the elevation of the upper limb (Roberts, 1974; Corruccini and Ciochon, 1976; Larson and Stern, 1986; Larson, 1993). While in African apes the supraspinatus is relatively broader, especially in gorillas (Schultz, 1930; Inman *et al*, 1944; Raven, 1950; Roberts, 1974), in orangutans and modern humans the proportions of the supraspinatus are only slightly larger than those seen in quadrupeds, i.e. the muscle is long and narrow compared to the infraspinatus (Ashton and Oxnard, 1963).

Electromyographic analyses provided additional information confirming a great part of the conclusions that were previously obtained from osteological data (Inman *et al*, 1944; Jungers and Stern, 1984; Tuttle and Basmajian, 1978a, 1978b; Larson and Stern, 1986, 1987), as well analyses done on the muscular weights (Potau *et al*, 2009). In general, the supraspinatus seems to play a similar role in different species of primates, depending on their locomotor behaviors. That is, the supraspinatus is extremely important in quadrupedal postures and locomotor support phases, as it stabilizes the glenohumeral joint; it plays a major role assisting the deltoid during the abduction of the upper limb in free arm movements; and it may also occasionally make a small contribution during the swing phase of the quadrupedal locomotion (Larson and Stern, 1986, 1987; Larson, 1989).

The knuckle-walkers are the hominoids that possess relatively broader scapulae and large supraspinous fossae, i.e. gorillas and chimpanzees (and a large infraspinous fossa, as we shall see later; Schultz, 1930, Roberts, 1974; Aiello and Dean, 1990; Potau *et al*, 2007; Young, 2008). In *Gorilla*, the most highly adapted hominoid for quadrupedalism, the supraspinous fossa is relatively larger than in other primate species; this can be explained because these apes need a greater stabilization due to their extremely large body size and the compressive forces that they have to bear since they use the upper limb in an abducted and externally rotated position during feeding (Roberts, 1974; Taylor and Slice, 2005; Potau *et al*, 2007). This has been confirmed by electromyographic studies that showed that during knuckle-walking, the supraspinatus, together with the infraspinatus, are essential stabilizers of the glenohumeral joint, which undergoes severe shearing stress, controlling specifically the humeral retraction during the support phase (Roberts, 1974; Tuttle and Basmajian, 1978b; Larson and Stern, 1987; Larson and Stern, 1992). During the first part of the swing-phase of the arm-rising the supraspinatus is recruited in concert with the deltoid muscle to abduct the upper limb in hominoids (Inman *et al*, 1944; Tuttle and Basmajian, 1978a; Larson and Stern, 1986; Larson and Stern, 1989). This is especially important during brachiation, vertical climbing and suspensory locomotion (Inman *et al*, 1944; Oxnard, 1966, 1967; Ashton and Oxnard, 1964; Ciochon and Corruccini, 1977). Apart its role as abductor, the supraspinatus, with its attachment on the top of the great tubercle, bears the main responsibility for preventing the superior displacement of the humeral head provoked by the deltoid and for stabilizing the humerus during the initial phase of brachial elevation (Howell *et al*, 1986; Larson and Stern, 1986; Thompson *et al*, 1996). Once the initial phase of brachial elevation passes, the tendency of the deltoid to displace the humerus decreases, and then the supraspinatus acts to assist the deltoid in the abduction (Larson and Stern, 1986).

In orangutans the supraspinous fossa may seem poorly developed due to the broad infraspinous, but it is, in fact, better developed than in any pronograde primate (Roberts, 1974) although it is quite small if compared to the African apes (Oxnard, 1984; Young, 2002). This supraspinous fossa, combined with the relatively large infraspinous fossa, may reflect the great range of circumduction that the orangutan can display when it has the forelimb raised above the shoulder level, being necessary a much greater muscular force to secure the shoulder joint (Roberts, 1974). However the especial locomotor behaviour of orangutans is reflected in a scapula that shares some similarities with that of the ateline monkey *Lagothrix* (Young, 2003) and of other arboreal quadruped primates, as will be seen below. Hylobatids are different from the great apes because they have a relatively small supraspinous fossa (and, as we shall see later, a relatively small infraspinous fossa), a difference that can reflect the overwhelming lineage-specific specializations of their shoulder for brachiation (Young, 2008). Hylobatid dissimilarities to the great apes, and the specialized nature of their locomotion, i.e., a true brachiation, are considered by some authors to reflect specialized and divergent adaptations; so, the great apes are more likely to reflect the ancestral morphotype, being less specialized than hylobatids (Young, 2003). In modern humans the habitual use of the arm below the level of the shoulder and the fact that the forelimb has mainly manipulative skills, reduced the levels of shearing stress in the joint and thus the need for a developed supraspinatus muscle, being the supraspinous fossa as well as the rest of the rotator cuff muscles relatively poorly developed (Roberts, 1974; Potau *et al*, 2007). This reduction of the rotator cuff is beneficial for modern humans as it provides to the glenohumeral joint greater speed and precision on its movements, which is essential to the manipulative skills of the modern human upper limb (Potau *et al*, 2007). Electromyographic analyses of modern humans confirm that the supraspinatus acts together with the deltoid to abduct the arm (Inman *et al*, 1944). Thus, we find in humans a unique arrangement of the shoulder muscles. Some of their features are similar to those of apes, e.g. their extremely huge deltoid, while others resemble those of quadrupeds, e.g. the reduction of the development of the rotator cuff muscles. This particular mosaic distribution allows modern humans to raise the arm as freely as that of apes, but without their power, duration and frequency (Ashton and Oxnard, 1963, 1964; Oxnard, 1967, 1969; Larson, 1993). Few studies have examined the molecular characteristics of the rotator cuff muscles, and the few ones that were performed were related to the quantification and distribution of the muscular fibers (Singh *et al*, 2002; Schmidt and Schilling, 2007). Recently, we started to examine the mRNA expression of the different MHC (myosin heavy chain) isoforms of the rotator cuff muscles (Potau *et al*, 2011). Three different MHC isoforms can be found in the skeletal muscles of adult mammals, and their expression is related to the functional properties of the muscle fibres (Bottinelli and Reggiani, 2000). Those isoforms are the slow MHC-I, the fast MHC-IIa and the fastest MHC-IIx. A fourth isoform is expressed only in very small mammals, the MHC-IIb (Baldwin and Haddad, 2001). In general, the slow postural muscles express mainly the slow MHC-I isoform, with a variable expression of the MHC-IIa isoform (Fitts *et al*, 1991; Baldwin, 1996; Fitts and Widrick, 1996; Schiaffino and Reggiani, 1996; Rivero *et al*, 1999; Talmadge, 2000). Furthermore, in the powerful, fast, and less resistant muscles all three isoforms can be expressed in variable proportions, but with a predominance of the MHC-II isoforms (Harridge *et al*, 1998; Larsson and Moss, 1993). While pronograde primates show higher proportions of the slow and resistant MHC-I isoform, evidencing the importance of their supraspinatus as a postural muscle, hominoids show higher proportions of the two fast MHC-II isoforms than the slow MHC-I isoform, a typical expression of fast and powerful muscles, with low resistance to fatigue (Klitgaard *et al*, 1990; Harridge *et al*, 1996). This is an evidence for the elevator function of the supraspinatus in hominoids (Potau *et al*, 2011), as we have seen above. In addition, modern humans showed a higher expression of the fastest MHC-IIx than the rest of hominoids, which can be the result of the greater mobility and precision of the muscles of the arm to perform their manipulative function (Potau *et al*, 2011). We have to point out that, probably due to the small sample size, no differences were found within the expression patterns of the MHC isoforms of the different hominoids. Thus, our studies confirm that molecular techniques can effectively be useful to better understand the different locomotor behaviors displayed by primates.

Infraspinatus

As in the case of the supraspinatous fossa, the predominantly arboreal hominoids share a scapula with relatively broad infraspinous fossa (Frey, 1923; Schultz, 1930; Inman *et al*, 1944; Roberts, 1974; Corruccini and

Ciochon, 1976; Manaster, 1979; Shea, 1986; Larson, 1993). Within hominoids, African apes show the relatively broader infraspinous fossa, gorillas having a relatively broader infraspinous fossae than chimpanzee; this may be a reflection of the greater need for shoulder joint stabilization in gorillas (Schultz, 1930; Roberts, 1974; Corruccini and Ciochon, 1976; Larson, 1993). This is also in agreement with the conclusions obtained from analyses of the muscular weights (Potau *et al*, 2009) and from electromyographic analyses (Tuttle and Basmajian, 1978a, 1978b; Larson and Stern, 1987). During knuckle-walking, highly mobile forelimbs experience greater demands for a different type of glenohumeral joint stabilization, because the shoulder joint bears stress produced by the lateral orientation of the glenoid, resulting in a tendency of the humerus to be displaced dorsally (Roberts, 1974; Larson and Stern, 1987). Such a stabilization seems to be performed by the supraspinatus and the infraspinatus without the assistance of the other muscles of the rotator cuff (Tuttle and Basmajian, 1978b; Larson and Stern, 1987).

During pendant suspension and support phase of arm-swinging, the infraspinatus seems to play a special role in the stabilization of the glenohumeral joint, resisting transarticular stresses suffered by the shoulder joint providing the required humeral head depressor action, although other muscles may also contribute to maintain the joint integrity (Roberts, 1974; Larson and Stern, 1986). In addition, in motions combining abduction with lateral rotation (e.g. overhead reaches) the infraspinatus seems to act as the primary synergist of the deltoid (Larson and Stern, 1986; Larson, 1987), a role that can be explained by the superolateral orientation of the infraspinatus insertion facet on the greater tubercle of the humerus, instead of the more typical lateral orientation usually seen in pronograde primates (Larson and Stern, 1986; Larson, 1995). It therefore seems that the orientation and position of the infraspinatus insertion facet may be a useful indicator of the frequency of arm-raising and overhead forelimb postures (Larson, 1995): within hominoids, hylobatids display highly suspensory behaviors and have the more superolateral oriented facets, orangutans being in an intermediate position, and African apes and modern humans exhibiting the least superolateral oriented facets (Larson, 1995).

The infraspinous fossa of orangutans confirms the important role of the infraspinatus muscle in suspensory locomotion. Orangutans have a broad infraspinous fossa, i.e. in comparison to this fossa the supraspinous fossa appears to be poorly developed (Roberts, 1974). This broad infraspinous fossa is related to the quadrumanous climbing behavior of these Asian apes, and to the greater muscular force required to stabilize the glenohumeral joint while raising the forelimb above their heads (Roberts, 1974). The scapula of hylobatids differs from that of orangutans because the former exhibit a relatively narrower infraspinous fossa (probably related to their overwhelming specializations for true brachiation; see above). This difference can seem to be surprising due to the apparent similar locomotor behavior shared by Asian apes, but when one looks in detail to their behaviors one can see that the scapula of orangutans exhibits a mosaic of suspensory and quadrupedal features, being more closely related to arboreal quadrupeds despite their highly arboreal and suspensory behaviors (Young, 2008). Orangutans are much bigger than gibbons, and this is a crucial difference because they need to avoid lifting their body weight and thus combine both pronograde/orthograde suspensory locomotion and postures that require additional stabilization (Roberts, 1974; Young, 2008). It can thus be said that their principal type of locomotion is moving through the canopy using a slow, cautious pronograde suspension, with frequent "drop leaps" (Thorpe and Crompton, 2005; 2006). Another difference between hylobatids and orangutans concerns the orientation of the base of the scapular spine (Larson and Stern, 1986; Larson, 1995), a feature that seems to have a great influence in the line of action of the infraspinatus as a stabilizer during pendant suspension and the support phase of the arm-swinging. While hylobatids possess extremely acute angles, orangutans display very obtuse scapular spine angles (Larson, 1995), reflecting the different locomotor demands of these taxa.

In modern humans the infraspinatus seemingly participates in the regulation of a typical pendant position of the upper limb. Modern humans have a relatively wide infraspinous fossa (together with a relatively small supraspinous fossa; Roberts, 1974; Aiello and Dean, 1990) and have a nearly horizontal scapular spine, falling outside the range of all other primates; these features are related to the exclusive use of the forelimb as a manipulatory organ (Larson, 1995).

Subscapularis

During primate evolution the subscapularis has reduced its size, being the largest muscle of the rotator cuff in 'lower' primates and decreasing relatively, but only slightly, from phylogenetically plesiomorphic hominoids to modern humans (Inman *et al*, 1944). However, there was an evolutionary increase of the number of fascicules with the expansion of its origin, as well as an increase of the elongation of the scapula (this same skeletal change increased the infraspinous fossa; Inman *et al*, 1944; De Palma, 2008). The subscapularis displays internal functional differentiation, in terms of fiber orientation and origin from different regions of the subscapularis fossa (Larson and Stern, 1986). In the case better studied by Larson and Stern (1986, 1987), the chimpanzees, three different parts are distinguished, and similar structures seem to be found in the other hominoids: there is an upper portion arising in the upper region of the subscapularis muscle, approximately opposite to the supraspinous fossa; the middle portion arises opposite to the infraspinous fossa, but in the anterior side of the scapula; and the lower portion arises from the ventral sulcus, lateral to the axillary bar, and from a common fascial plane with the teres major muscle (Larson and Stern, 1986). Each differentiated region of the muscle acts individually, depending on arm position, to help controlling the rotator state of the humerus (Larson and Stern, 1986; Larson, 1988). Moreover, there seems to be a relationship between the region of the facet of the lesser tubercle of the humeral head, where the fibers insert, and the function of those specific fibers (Larson, 1995). Thus, the abduction and medial rotation is carried out by those fibers inserting onto the more proximal region of the facet, corresponding to the upper portion of the subscapularis, while the adduction and medial rotation is conducted by those fibers inserting more distally, corresponding to the lower portion of the subscapularis (Larson and Stern, 1986; Larson, 1988; Larson, 1995).

The insertion facets in the lesser tubercle of the humerus are thought to be a useful tool to predict the degree of functional differenciation within the subscapularis (Larson, 1995). Hominoids have the longest and narrowest subscapularis insertion facets, confering a greater degree of versatility of the muscle for influencing the position and motion of the humeral head (Larson, 1995); this is important in order to provide higher mobility to the shoulder joint. In summary, the subscapularis is mainly a medial rotator of the arm, rotation that can be combined with abduction (by the upper portion) or adduction (by the lower portion) depending on the position and movements of the upper limb (Larson and Stern, 1986; Larson, 1988). The medial rotation of the humerus is especially important during knuckle-walking in African apes, as it contributes to stabilize the glenohumeral joint (Tuttle and Basmajian, 1978b; Larson and Stern, 1987), compensating for the shearing force that this joint bears due to the dorsal position of the scapula (Roberts, 1974; Larson, 1993). This is consistent with the results obtained from eletromyographic studies (Tuttle and Basmajian, 1978b; Larson and Stern, 1987) and the proportionally large mass of the subscapularis found in chimpanzees (Inman *et al*, 1944; Potau *et al*, 2009).

During the support phase of the arm-swinging of chimpanzees, the subscapularis has no contribution to joint stabilization (Larson and Stern, 1986; Larson, 1988), demonstrating that the force couple model of Inman *et al* (1944) was wrong (see above). By contrast, this muscle is recruited during the support phase of the arm-swinging in hylobatids (Jungers and Stern, 1984), a difference that can be interpreted as a rotatory function, rather than a stabilizing function, because in hylobatids there is a greater participation in free arm movements than in chimpanzees (Larson, 1988). The most important function of the subscapularis takes place during the support (or "pull-up") phase of vertical climbing, where its role as medial rotator becomes extremely important (Larson and Stern, 1986; Larson, 1988). This is a major difference between the subscapularis and the other rotator cuff muscles, and another evidence of the individuality of these muscles. Climbing species are thus expected to have larger subscapularis muscles (Larson, 1988), associated with an increased enlarging of the lateral expansion of the subscapular fossa beyond the ventral bar (the area corresponding to the origin of the lower subscapularis; Larson, 1988; Larson, 1995; Larson and Stern, 1986).

As predicted, great apes show a relatively wider lateral expansion of the subscapular fossa, but not as wide as that found in hylobatids (Larson, 1995). This greater contribution of the subscapularis to the shoulder motion in hylobatids is related to the low degree of humeral head torsion, and the resulting "lateral set" to the gibbon's elbow joint (Larson, 1988), considered an adaptive advantage for limb positioning during brachiation. The "lateral set" refers to the posture of the limbs that these animals adopt while they are resting, with the long axis of the humeral head perpendicular to the glenoid fossa, and the elbow joint in a position such as the arm seems to be laterally rotated, with their forearms splayed out to the side (Larson, 1988). In other words, it is the elbow that is rotated, and not the head of the humerus as it may seem at first sight. The increased demands on the susbcapularis can be explained to overcome the "lateral set", that is, the lateral rotation at the elbow, during free arm movements by the medial rotation of the muscle (Larson, 1995), in order to reestablish the correct arm position needed for normal reaching motions (Larson, 1988). The limited humeral torsion seen in hylobatids can be a compromise between the need to maintain a transverse axis at the elbow joint, which is a consequence of the scapular dorsal reorientation, and the demand for an extreme positioning of the elbow during arm-swinging (Larson, 1988). In the case of modern humans, they do not frequently use their forelimbs for climbing, so the lateral expansion of the subscapularis fossa is relatively small in relation to their body size (Larson, 1995).

Teres Minor

The teres minor is a muscle that is absent in phylogenetically plesiomorphic mammals, being derived from the deltoid complex (Inman *et al*, 1944; De Palma, 2008). When the infraspinous fossa started to progressively increase, a part of the deltoid separated from it, and started to increase too. Over the time, the skeletal changes have altered the position of this "new" muscle, permitting it to act upon the humerus in a downward direction (Inman *et al*, 1944; De Palma, 2008). According to Ashton and Oxnard (1963), in hominoids the teres minor is completely differentiated from the infraspinatus and deltoid muscles to form a separate unit. For a long time the teres minor was thought to form a functional group with the subscapularis and the infraspinatus (see above), depressing the head of the humerus and rotating it (Inman *et al*, 1944; Tuttle and Basmajian, 1978a, 1978b). However, we currently know that in general terms the teres minor is primarily a lateral rotator of the humerus, participating in the maintenance of the humeral head in position (Larson and Stern, 1986; Aiello and Dean, 1990; Jenkins, 2009).

It is important to highlight that the teres minor and, as we have seen above, a great part of the subscapularis, are not involved in the swing phase of the arm-raising (Larson and Stern, 1986), being only active during its support phase. This may indicate that, at least in arm-raising behaviors, teres minor helps in resisting transarticular tensile stresses at the glenohumeral joint (Larson and Stern, 1986), and as a retractor of the humerus during hoisting in orangutans (Tuttle and Basmajian, 1987a). That is, it is an adductor of the arm (Larson and Stern, 1986). During knuckle-walking, teres minor acts, at least in chimpanzee, in the rotation of the arm laterally (Larson and Stern, 1987). From this review we can conclude that of the muscles of the rotator cuff, the teres minor is the least studied and further studies are needed to better understand its function and evolution.

Conclusion

Understanding primate and human rotator cuff anatomy is crucial for enhancing our understanding of primate upper limb adaptations and thus of primate and human evolutionary history. Each muscle of the rotator cuff has its own role in the movement of the upper extremity, this role varying in different locomotor groups and being adapted to the functional requirements of each type of locomotion. Thus, if we understand how the rotator cuff muscles work in each locomotor groups, we can improve our understanding about which were the evolutionary processes that took place in the origin and evolution of the hominoids and thus of our lineage.

Acknowledgments

We would like to thank the following curators that kindly provided access to the specimens under their care: Georges Lenglet (Royal Belgian Institute of Natural Sciences) and Eulàlia García (Natural History Museum of Barcelona). This research was funded in part by the Spanish Ministry of Education and Science (MEC) Funding for Research Projects (MEC CGL2010), by the Generalitat de Catalunya Funding for Consolidate Research Groups (DURSI 2009SGR-00884), and by the Predoctoral Fellowship Grant Program of the Ministry of Education and Science (MEC) FPU (AP2008-00877) and its research travel grants, granted to GBH.

References

Aiello, L; Dean, C (1990) *An introduction to human evolutionary anatomy*. Academic Press, London.

Ashton, EH; Oxnard, CE (1963) The musculature of the primate shoulder. *Trans Zool Soc Lond,* 29: 553-650.

Ashton, EH; Oxnard, CE (1964) Functional adaptations in the primate shoulder girdle. *Proc Zool Soc Lond,* 142: 49-66.

Baldwin, KM (1996) Effects of altered loading stated on muscle plasticity: What have we learned from rodents? *Med Sci Sports Exer,* 28: 101-106.

Baldwin, KM; Haddad, F (2001) Plasticity in skeletal, cardiac, and smooth muscle. *J App Phys,* 90: 345-357.

Botinelli, R; Reggiani, C (2000) Human skeletal muscle fibres: Molecular and functional diversity. *Prog Biophys Mol Biol,* 73: 195-262.

Ciochon, RL; Corruccini, RS (1977) The coraco-acromial ligament and projection index in man and other anthropoid primates, *J Anat,* 124: 627-632.

Corrucini, RS; Ciochon, RL (1976) Morphometric affinities of the human shoulder. *Am J Phys Anthrop,* 45: 19-38.

De Palma, AF (2008) The origin and comparative anatomy of the pectoral limb. *Clin Orthop Relat Res,* 466: 531-542.

Doran, DM (1993) Comparative locomotor behavior of chimpanzees and bonobos: The influence of morphology on locomotion. *Am J Phys Anthrop,* 91: 83-98.

Doran, DM; Hunt, KD (1994) The comparative locomotor behavior of chimpanzees and bonobos: Species and habitat differences. IN: *Chimpanzee cultures* (eds. RW Wrangham, WC McGrew, FBM de Waal, PG Heltne), pp. 93-108. Cambridge, MA: Harvard University Press.

Fitts, RH; McDonald, KS; Schluter, JM (1991) The determinants of skeletal muscle force and power: Their adaptability with changes in activity pattern. *J Biomech,* 24: 111-122.

Fitts, RH; Widrick, JJ (1996) Muscle mechanics: Adaptations with exercise-training. *Ex Sports Sci Rev,* 24: 427-473.

Fleagle, JG (1999) *Primate adaptation and evolution* (2nd ed).Academic Press, London.

Frey, H (1923) Untersuchungen über die scapula, speciell über ihre äussere form und deren abhängigkeit von der function. *Zeitschr. f. Anat.* U. Entwicklungsgesch, 68: 277-324

Gebo, DL (2010) Locomotor function across Primates (including humans). In: *A companion to biological anthropology* (ed. CS Larsen), pp. 530-544. Chichester, UK; Malden, MA: Wiley-Blackwell.

Goodall, J (1965) Chimpanzees of the Gombe Stream Reserve. In: *Primate behavior* (ed. I DeVore), pp. 425-473. Hort Rinehart and Winston.

Green, DJ; Hamrick, MW; Richmond, BG (2011) The effects of hypermuscularity on shoulder morphology in myostatin-deficient mice. *J Anat,* Vol. 218 (5): 544-557.

Harridge, SD; Bottinelli, R; Canepari, M; Pellegrino, M; Reggiani, C; Esbjornsson, M, et al. (1996) Whole muscle and single fibre contractile properties and myosine isoforms in humans. *Pflügers Archiv,* 432: 913-920.

Harridge, SD; Bottinelli, R; Canepari, M; Pellegrino, M; Reggiani, C; Esbjornsson, M, et al. (1998) Sprint training, in vitro and in vivo muscle function, and myosin heavy chain expression. *J App Phys*, 84: 442-449.

Hepburn, D (1892) The comparative anatomy of the muscles and nerves of the superior and inferior extremities of the anthropoid apes. *J Anat Lond,* 26: 149 and 324.

Hess, SA (2000) Functional stability of the glenohumeral joint. *Man Ther,* Vol. 5(2): 63-71.

Howell, JA (1917) An experimental study of the effect of stress and strain on bone development. *Anat Rec,* 3: 233-252.

Howell, SM; Imobersteg, AM; Seger, DH; Marone, PJ (1986) Clarification of the role of the supraspinatus muscle in shoulder function. *J Bone Joint Surg,* Vol. 68 (3): 398-404.

Inman, VT; Saunders, JB; Abbott, LC (1944) Observations on the function of the shoulder joint. *J Bone Joint Surg,* Vol. XXVI, No 1: 1-30.

Jenkins, DB (2009) *Hollinshead's functional anatomy of the limbs and back* (9th ed). Saunders, Elsevier.

Jungers, WL; Stern, JT (1984) Kinesiological aspects of brachiation in lar gibbons. In: *The lesser apes* (eds. H Preuschoft, DJ Chivers, WY Brockelman & N Creel), pp. 119-134. Edinburgh University Press.

Keith, A (1923) *Man's posture: Its evolution and disorders.* British Med J, n°1.

Klitgaard, H; Mantoni, M; Schiaffino, S; Ausono, S; Gorza, L; Laurent-Winter, C, et al. (1990) Function, morphology and protein expression of agein skeletal muscle: A cross-sectional study of eldery men with different training backgrounds. *Acta Physiol Scan,* 140: 41-54.

Larson, SG (1988) Subscapularis function in gibbons and chimpanzees: implications for interpretation of humeral head torsion in hominoids. *Am J Phys Anthrop,* 76: 449-462.

Larson, SG (1993) Functional morphology of the shoulder in Primates. In: *Postcranial adaptation in nonhuman primates* (ed. DL Gebo), pp. 45-69. DeKalb: Northern Illinois University Press.

Larson, SG (1995) New characters for the functional interpretation of primate scapulae and proximal humeri. *Am J Phys Anthrop,* 98: 13-35.

Larson, SG; Stern, JT (1986) EMG of scapulohumeral muscles in chimpanzee during reaching and "arboreal" locomotion. *Am J Anat,* 176: 171-190.

Larson, SG; Stern, JT (1987) EMG of chimpanzee shoulder muscles during knuckle-walking: problems of terrestrial locomotion in a suspensory adapted primate. *J Zool Lond,* 212, 629-655.

Larson, SG; Stern, JT (1989) The role of suprapsinatus in the quadrupedal locomotion of vervets (Cercopithecus aethiops): Implications for interpretation of humeral morphology. *Am J Phys Anthrop,* 79: 369-377.

Larson, SG; Stern, JT (1992) Further evidence for the role of supraspinatus in quadrupedal monkeys. *Am J Phys Anthrop,* 87: 359-363.

Larsson, L; Moss, RL (1993) Maximum velocity of shortening in relation to myosin isoform composition in single fibres from human skeletal muscles. *J Physiol*, 472: 595-614.

Le Gros Clark, WE (1959) *The antecedents of man: An introduction to the evolution of Primates.* Edimburg University Press.

Napier, JR (1963) Brachiation and brachiators. In: The Primates (eds. JR Napier & NA Barnicot). *Symp Zool Soc Lond.* N°10: 183-194.

Oxnard, CE (1966) *Some functional osteometric features of the primate innominate bone.* Proc Int Primat Soc Franckfurt.

Oxnard, CE (1967) The functional morphology of the primate shoulder as revealed by comparative anatomical, osteometric and discriminant function techniques. *Am J Phys Anthrop,* 26: 219-240.

Oxnard, CE (1969) Evolution of the human shoulder: some possible pathways. *Am J Phys Anthrop,* 30: 319-332.

Oxnard, CE (1984) *The order of man: a biomathematical anatomy of the primates.* New Haven: Yale University Press.

Oxnard, CE; Neely, PM (1969) The descriptive use of neighbourhood limited classification in functional morphology: An analysis of the shoulder in primates. *J Morphol,* 129: 127-148.

Potau, JM; Artells, R; Bello, G; Muñoz, C; Monzó, M; Pastor, JF; de Paz, F; Barbosa, M; Diogo, R; Wood, B (2011) Expression of myosin heavy chain isoforms in the supraspinatus muscle of different primate species: implications for the study of the adaptation of primate shoulder muscles to different locomotor modes. *Int J Prim,* Vol. 32(4): 931-944.

Potau, JM; Bardina, X; Ciurana, N (2007) Subacromial space in African great apes and subacromial impingement syndrome in humans. *Int J Primatol,* 28: 865-880.

Potau, JM; Bardina, X; Ciurana, N; Camprubí, D; Pastor, JF (2009) Quantitative analysis of the deltoid and rotator cuff muscles in humans and great apes. *Int J Primatol,* 30: 697-708.

Raven, HC (1950) *The anatomy of the gorilla.* New York, Columbia University Press.

Remis, M (1995) Effects of body size and social context on the arboreal activities of lowland gorillas in the Central African Republic. *Am J Phys Anthrop,* 97 (4): 413-433.

Rivero, JL; Talmadge, RJ; Edgerton, VR (1999) Interrelationships of myofibrillar ATPase activity and metabolic properties of myosin heavy chain-based fibre types in rat skeletal muscle. *Histochem Cell Biol,* 111: 277-287.

Roberts, D (1974) Structure and function of the primate scapula. En: *Primate locomotion* (ed. FA Jenkins), pp. 171-200. Academic Press, New York.

Schiaffino, S; Reggiani, C (1996) Molecular diversity of myofibrillar proteins: Gene regulation and functional significance. *Physiol Rev,* 76: 371-423.

Schmidt, M; Schilling, N (2007) Fiber type distribution in the shoulder muscles of the tree shrew, the cotton-top tamarin, and the squirrel monkey related to shoulder movements and forelimb loading. *J Hum Evol,* 52: 401-419.

Schultz, AH (1930) The skeleton of the trunk and limbs of higher primates. *H Biol,* Vol. 2 nº3: 303-438.

Shea, BT (1986) Scapular form and locomotion in chimpanzee evolution. *Am J Phys Anthrop,* 70: 475-488.

Singh, K; Melis, EH; Richmond, FJR; Scott, SH (2002) Morphometry of Macaca mulatta forelimb. II. Fiber-type composition in shoulder and elbow muscles. *J Morph,* 251: 323-332.

Sonnabend, DH; Young, AA (2009) Comparative anatomy of the rotator cuff. *J Bone Joint Surg* (Br), Vol. 91B: 1632-1637.

Susman, RL (1984) *The pygmy chimpanzee: Evolutionary biology and behaviour.* New York: Plenum Press.

Talmadge, RJ (2000) *Myosin heavy chain isoform expression following reduced neuromuscular activity: Potential regulatory mechanisms.* Muscle & Nerve, 23: 661-679.

Taylor, AB; Slice, DE (2005) A geometric morphometric assessment of the relationship between variation and locomotion in African Apes. In: *Modern morphometrics in physical anthropology* (ed. DE Slice). Kluwer Academic/Plenum Publishers, New York.

Terry, GC; Chopp, TM (2000) Functional anatomy of the shoulder. *J Athl Train,* 35 (3): 248-255.

Thompson, WO; Debski, RE; Boardman, ND; Taskiran, E; Warner, JJP; Fu, FH; Woo, SLY (1996) A biomechanical analysis of rotator cuff deficiency in cadaveric model. *Am J Sports Med,* Vol.24 (3): 286-292.

Thorpe, SKS; Crompton, RH (2005) Locomotor ecology of wild orangutans (Pongo pygmaeus abelii) in the Gunung Leuser ecosystem, Sumatra, Indonesia: A multivariate analysis using log-linear modelling. *Am J Phys Anthrop,* 127: 58-78.

Thorpe, SKS; Crompton, RH (2006) Orangutan positional behavior and the nature of arboreal locomotion in Hominoidea. *Am J Phys Anthrop,* 131: 384-401.

Tuttle, RH (1975) Parallelism, brachiation, and hominoid phylogeny. In: *Phylogeny of the Primates* (eds. WP Luckett & FS Szalay), pp. 447-480. New York: Plenum Press.

Tuttle, RH; Basmajian, JV (1978a) Electromyography of pongid shoulder muscles II: deltoid, rhomboid and "rotator cuff". *Am J Phys Anthrop,* 49: 47-56.

Tuttle, RH; Basmajian, JV (1978b) Electromyography of pongid shoulder muscles III: quadrupedal positional behavior. *Am J Phys Anthrop,* 49: 57-70.

Tuttle, RH; Watts, DR (1985) The positional behaviour and adaptive complexes of Pan gorilla. In: *Primate morphophysiology, locomotor analysis and human bipedalism* (ed. S Kondo), pp. 261-288. University of Tokyo Press, Tokyo.

Ward, CV (2007) Postcranial and locomotor adaptations of hominoids. In: *Handbook of Paleoanthropology*. Part II (eds. W Henke & I Tattersall), pp. 1011-1030. Springer-Verlag Berlin Heidelberg New York.

Ward, SR; Hentzen, ER; Smallwood, LH; Eastlack, RK; Burns, KA; Fithian, DC; Friden, J; Lieber, RL (2006) Rotator cuff muscle architecture. *Clin Orth Rel Res,* 448: 157-163.

Wolffson, DM (1950) Scapula shape and muscle function, with special reference to the vertebral border. *Am J Phys Anthrop,* 8: 331-341.

Young, NM (2002) *Homology and homoplasy in the hominoid postcranium.* PhD dissertation. Harvard University.

Young, NM (2003) A reassessment of living hominoid postcranial variability: implication for ape evolution. *J Hum Evol,* 45: 441-464.

Young, NM (2008) A comparison of the ontogeny of shape variation in the Anthropoid scapula: functional and phylogenetic signal. *Am J Phys Anthrop,* 136: 247-264.

Ziegler, AC (1964) Brachiating adaptations of chimpanzee upper limb musculature. *Am J Phys Anthrop,* 22: 15-32.

In: Primates
Editors: E. F. Hughes and M. E. Hill
ISBN: 978-1-62081-498-7

Chapter IV

Neurobiological Correlates of Behavioral and Cognitive Performance in Nonhuman Primates*

***Gwendolen E. Haley*[1,2] *and Jacob Raber*[1,2,3]**
[1]Behavioral Neuroscience Department, Oregon Health and Science University Portland, OR, US
[2]Division of Neuroscience,
Oregon National Primate Research Center, Beaverton, OR, US
[3]Department of Neurology,
Oregon Health and Science University, Portland, OR, US

Abstract

Animals in experiments are traditionally grouped by experimental treatment. Although this is a valuable way to differentiate the groups, alternatively, groups can be distinguished based on cognitive performance. Performance based analysis can yield valuable insights, corresponding to behavior and/or cognition, that might not otherwise be

* A version of this chapter also appears in Monkeys: Biology, Behavior and Disorders, edited by Rachel M. Williams, published by Nova Science Publishers, Inc. It was submitted for appropriate modifications in an effort to encourage wider dissemination of research.

observed. As an example of such an analysis, we discuss a cohort of elderly female rhesus macaques who participated in a spatial food port maze navigational test. Circadian activity and pharmacological MRI (phMRI) were assessed in these monkeys in vivo and radioligand binding was assessed in post-mortem tissue. Based on cognitive performance in the spatial maze, the cohort of monkeys was divided into Good Spatial Performers (GSP) and Poor Spatial Performers (PSP). GSP animals were more active during the day and less active at night compared to PSP animals. In addition, GSP animals had a higher percentage change in blood-oxygen-level-dependent (BOLD) signal after a scopolamine challenge, a non-specific muscarinic receptor antagonist, compared to PSP animals. Post-mortem radioligand experiments demonstrated that hippocampal muscarinic type 1 (M1) maximum receptor binding and receptor binding affinity, hypothesized to have an integral role in spatial learning and memory, was significantly greater in the hippocampus of GSP than PSP animals. In contrast to the hippocampus, M1 receptor binding was not significantly different in the prefrontal cortex (PFC). Muscarinic type 2 (M2) maximum receptor binding and receptor binding affinity were not significantly different between the two groups in either brain region. Finally, there were positive correlations between circadian measures and the percentage change in BOLD signal following the scopolamine challenge, as well as M1 receptor binding measures. Thus, GSP animals sleep more and have anenhanced M1 receptor function. These data demonstrate the close relationship between BOLD signal changes, circadian activity, and M1 receptor binding parameters. Distinguishing groups based on cognitive or behavioral performance is valuable for studying neurobiological correlates of performance in nonhuman primates.

Keywords: Behavioral groups, muscarinic receptors, phMRI

Introduction

Traditionally, when designing experiments, animals are grouped based on an experimental manipulation. During the course of the experimentation, especially when assessing animal behavior or cognition, a variety of data is collected. An alternative way to examine group differences is a post-hoc type of analysis in which animals are grouped based on their behavioral or cognitive performance. For instance, post-mortem tissue analyses could conceivably present different results that could be related to a specific type of behavior or cognitive testing rather than the experimental condition. This approach can especially be successful when using a multidisciplinary approach to examine the animals. In this chapter, we will present and discuss such analysis using data from a previous study in which muscarinic receptor function was assessed *in vivo* and *in vitro* in post-mortem tissue from the same animals. Traditional experiments separate groups based on an independent variable. For instance, when assessing age-related changes in a particular physiological area, distinct groups are traditionally based on age [1]. Likewise, groups can be based on genetic differences [2]. Alternatively, groups can be based on a specific treatment versus control [3, 4]. However, with behavioral or cognitive experiments, performance-based groups might demonstrate neurobiological differences with strong implications in translational research projects. As invasive diagnostic methods are used sparingly in the clinic, an idea of what physiological parameters are reflected in behaviors would be most valuable. Specifically, with regard to aging and age-related pathologies, understanding behavioral and cognitive representation could improve health care for the increased aging population. Muscarinic acetylcholine receptors are associated with post-synaptic neurotransmission and memory function [5]. They are present at high concentrations in the prefrontal cortex (PFC) and the hippocampus [6] and have an integral role in spatial learning and memory in rodents, nonhuman primates, and humans [7-10]. Scopolamine, a nonspecific muscarinic receptor antagonist [11], impairs cognitive task performance in rats [12], dogs [13], rhesus monkeys [14, 15], and humans [16]. The most highly expressed muscarinic receptor in the PFC and hippocampus, regions critical for cognitive function, is the type 1 (M_1) subtype [7-10, 17-19]. The muscarinic type 2 (M_2) is the second most abundant muscarinic receptor in the PFC and hippocampus [20, 21] and might also contribute to cognitive function [22]. M_1 receptor deficient mice show impaired working memory and memory consolidation and M_2 receptor deficient mice show impaired passive avoidance task performance [23]. Nonhuman primate studies show a decrease in M_1 and M_2 receptor number with age [24]. Other reports indicate an age-related decrease in M_1 receptor affinity [25]. In patients with Alzheimer's disease, M_1 receptor agonists are proposed to increase cognitive function, increase neurogenesis, and decrease tau phosphorylation [26-29]. Together, these findings suggest that the M_1, and to a lesser extent M_2, receptors may play a key role in the regulation of cognitive function in the elderly. Cognitive function is negatively impacted by disruptions in circadian activity, especially in the elderly [30, 31]. For example, circadian activity has been correlated with cognitive performance on a spatial task in old female rhesus monkeys [32]. The M_1 receptor might be involved in regulation of circadian activity [33, 34] and this in turn might contribute to the M_1 receptor-mediated cognitive effects. Neurotransmitter function can be readily assessed *in vivo* using pharmacological magnetic resonance imaging (phMRI). Although similar to functional magnetic resonance imaging (fMRI), which measures blood oxygen level dependent (BOLD) signal changes associated with cognitive performance, phMRI measures BOLD signal changes following exposure to pharmacological stimuli [35]. This technique has been successfully used in humans and nonhuman primates to study the function of the dopamine system [36]. For example, in experimentally 1-methyl-4-phenyl-1,2,3,6-tetrahydropyridine (MPTP)-lesioned rhesus monkeys, the BOLD signal was found to be positively correlated with the number of surviving dopaminergic neurons [37]. Importantly, because the response of these neurons to dopaminergic agonists is lesion-dependent, phMRI can be utilized clinically to assess the efficacy of therapies in Parkinson's disease patients [38]. Although fewer phMRI studies have focused on the acetylcholine system [39, 40], it is plausible that they also will yield clinical applications.

Studies Discussed in this Chapter

In the present chapter, we discuss studies in which muscarinic receptor function was assessed in a cohort of cognitively tested elderly female rhesus macaques *in vivo* using phMRI and *in vitro* using saturation binding assays.

Based on the number of trials to reach a criterion performance in the food port spatial maze navigational test [32], the monkeys were classified into two groups: good spatial performers (GSP) and poor spatial performers (PSP). In the phMRI experiment, the animals were given scopolamine and BOLD signal changes in the PFC and hippocampus were assessed. Subsequently, postmortem M_1 and M_2 receptor binding assays were performed using the PFC and hippocampus tissues of these animals. Finally, the potential relationships between circadian activity and the percentage change in BOLD signal and number and binding affinity of M_1 and M_2 receptors were explored.

Nonhuman Primate Spatial Maze

A nonhuman primate spatial maze was developed using a cohort of elderly female rhesus macaques [32]. In this maze, a bank of 10 food ports was mounted to the wall (Figure 1). Each port consisted of a one-way opaque swinging door. Freely moving monkeys were placed in the room and were allowed to search the ports. One of the 10 food ports was repeated baited with preferred candy until the animals searched the port on the first or second search, three times in a row (location 1).

Once the monkeys learned the first location, the baited food port was switched to a different one (location 2). Again, the monkeys were allowed to search the series of food ports until they searched the baited port on the first or second port searched three consecutive times.

The number of trials to criterion was counted as well as the search strategies. Monkeys that used a spatial search (searching within 2 ports of the baited port) completed the task in fewer trials than the monkeys that utilized a serial search pattern (searching the ports consecutively).

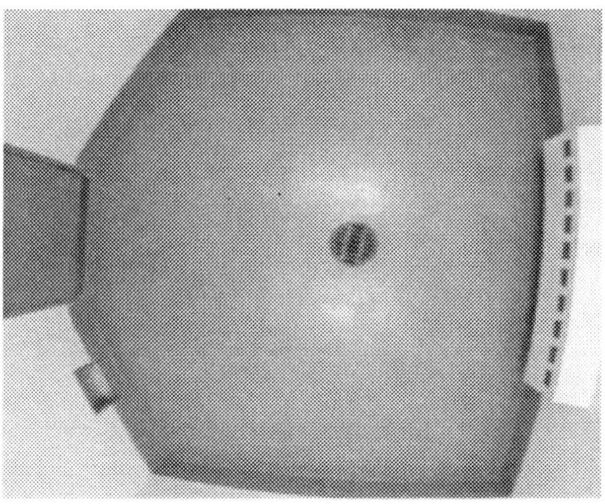

Figure 1. Image of the Spatial Foodport Maze Room. The room is 2.44 x 3.45 m and consists of a human entry door (equipped with a one-way observation window), animal entry door, one-way observation window (not visible in the image), and a custom made 157 x 38 cm bank of 10 ports. The port apparatus is located on the wall opposite the entry doors. Each port measuring 7.6 x 10.2 x 7.6 cm has a one-way opaque swinging door with 5 cm between each port. Cameras were mounted in recesses in the ceiling to record animal behavior in the room and during the task.

Circadian Activity

During the testing period, animals wore a nylon activity collar for 2 weeks. Day activity, night activity, number of times the animals woke during the night (wake bouts), and the time to fall asleep (sleep latency), were recorded and averaged over the 2 weeks.

Once the spatial maze was completed, the activity measures were correlated with performance on the spatial maze. Monkeys that performed better on the spatial maze had higher indices of sleep quality than those who performed poorly. Furthermore, the better performing monkeys were more active during the day than those who performed poorly [32]. After this task was completed, it appeared that there were 2 separate groups of monkeys; one that performed well on the task and one that did not.

Behavioral or Cognitive Classification

Based on the performance of the cohort in the spatial maze, individual monkeys were separated into two groups: good spatial performers (GSP) and poor spatial performers (PSP). GSP was defined as animals that completed the task in less than 30 trials in either the first or second position (18.7 ± 4.8 trials $n = 4$ first location; 7 and 14 trials $n = 2$ for the second location).

PSP was defined as animals that completed the task in more than 30 trials ($n = 6$; 81.8 ± 8.5 trials). Control animals for the phMRI experiment were animals from the same cohort that did not participate in the spatial maze testing ($n = 5$).

Although the number 30 might seem arbitrary, examining the data, there were monkeys that learned the first location very quickly (less than 30 trials). Or, if the monkeys did not learn the first location quickly, they were able to release the idea of the first location and learned the second location very quickly.

This observation is observed in the average number of trials in the first location and the two data points for the second location. Not every monkey exhibited the same pattern of learning. As for the PSP, using 30 trials as the cut-off was not detrimental to those monkeys in the PSP group, as the number of trials to criterion was much higher in the PSP compared to the GSP. There were no monkeys in the PSP group that completed the task in less than 40 trials to criterion.

There were 3 monkeys that were unable to complete the task due to temperament issue or lack of ability to be trained to search the ports. These monkeys would also be considered as PSP, however since there was no clear delineation, they were excluded from the remaining study. As previously mentioned, during the cognitive testing phase of the experiment, circadian activity was measured in the participating monkeys.

Table 1. Circadian activity levels of good spatial performers (GSP) and poor spatial performers (PSP)[1]

	GSP	PSP
Overall Activity	117.8 ± 13.0*	89.0 ± 5.8
Day Activity	213.0 ± 24.2*	143.2 ± 17.7
Night Activity	24.3 ± 2.3	22.3 ± 2.6
Light:Dark Activity Ratio	8.7 ± 0.8**	6.2 ± 0.5
Sleep Latency	21.8 ± 4.7	33.8 ± 5.1

[1]Units for Overall Activity, Day Activity, and Night Activity are average activity counts. Sleep latency is average time (min) to fall asleep. GSP (n = 6) compared to PSP (n = 6). *$P < 0.05$, **$P < 0.01$ GSP significantly greater than PSP.

The activity was monitored using the Actiwatch system for 2 weeks. Interestingly, we found that the monkeys that performed better on the spatial maze were more active during the day, less active at night, and took less time to fall asleep [32]. Further analysis demonstrated that GSP animals had significantly higher overall activity, higher day activity, and higher Light:Dark activity ratios than PSP animals (Table 1 [41]).

phMRI

Based on their performance in the spatial maze, we hypothesized that a performance difference observed in these monkeys was related to the cholinergic system, specifically the muscarinic receptor system as it has a major role in spatial learning and memory. Furthermore, studies suggest that the function of the acetylcholine system declines with age and is highly affected by Alzheimer's disease. A better understanding of how the muscarinic receptor functions among the different behavioral or cognitive groups of monkeys might yield valuable information on how neurobiology reflects behavior and cognition. To test this hypothesis, the cohort of monkeys from the spatial maze was first exposed to phMRI. In the phMRI, BOLD response to scopolamine, a non-specific muscarinic receptor antagonist, was measured. Monkeys were fasted the night before and the morning of MRI experiments. Monkeys were sedated transported from their home cages to the magnet located on the ONPRC campus. Animals were immediately intubated and respirated with isoflurane, and a MRI compatible catheter was inserted into the cephalic vein. The animal's head was placed inside an MRI coil and fitted with pads to eliminate head movement. A baseline scan of 20 contiguous T_2*-weighted multi-shot echo planar images were obtained over the course of 7 minutes (Figure 2). Immediately following the baseline scans, scopolamine (0.4 mg in 1 ml of isotonic saline or 1 ml of isotonic saline for control animals) was infused over 1 minute using an infusion pump. Following the intravenous infusion, a T_2-weighted and 2 T_1-weighted anatomical images were acquired, which was a total of 46 minutes, sufficient time for scopolamine to have CNS effects [42]. Immediately following the anatomical scans, a second BOLD series of T_2*-weighted images was acquired using the exact parameters as the baseline scan ("challenge scan"). The change in BOLD signal from baseline to challenge scan (scopolamine or vehicle) was determined. Isotonic saline control was used to ensure that the change observed between the baseline and challenge scan was due to a scopolamine effect, and not an artifact from signal drift of the magnet. Slice-dependent time shifts were interpolated, odd-even slice intensity differences removed, motion corrected, and T_1 and T_2 anatomical scans were co-registered. Baseline and challenge phMRI scans were averaged across the 20 volumes and the single image was normalized to the co-registered anatomical scan of each individual animal (PFC: Figure 3; hippocampus: Figure 4).

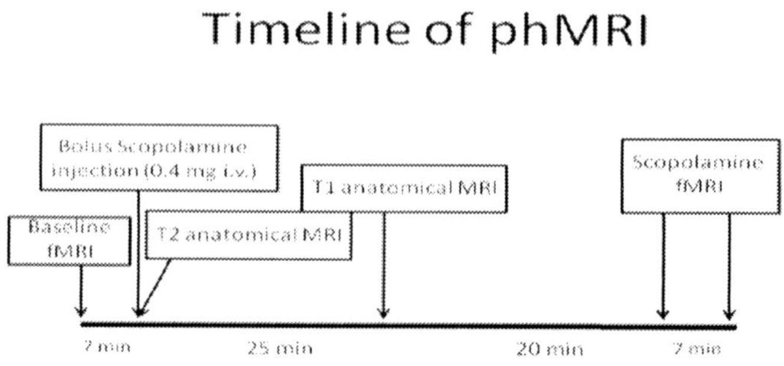

Figure 2. Timeline for pharmacological MRI (phMRI). Once animals were anesthetized, intubated, connected to physiological monitors, and placed in the scanner, a baseline multi-shot T_2* image series was collected. Immediately following the conclusion of the baseline scan, scopolamine was intravenously infused. During the 45 minute time period for the scopolamine to have action within the brain, anatomical images, both T_2 and two T_1 images, were acquired. At the conclusion of the anatomical scans, another scan with the exact parameters of the baseline scan was conducted and defined as a "challenge scan".

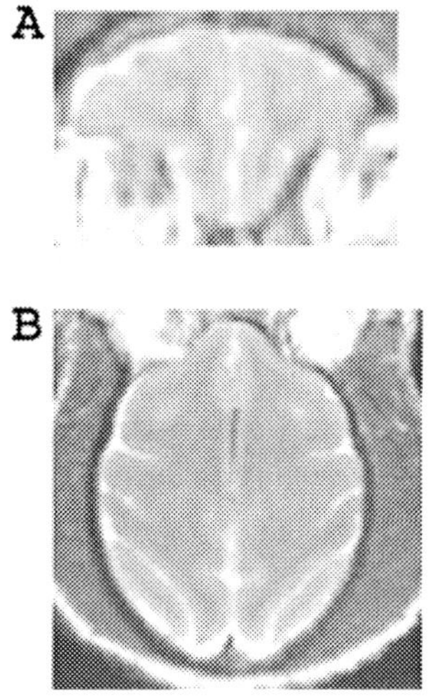

Figure 3. Representative structural MRI images from the prefrontal cortex (PFC) in A.coronal view and B.horizontal views. Regions of interest (ROIs) were drawn on structural images, using a functional overlay, to identify the areas of interest. Voxel intensity of the ROI in the GSP and PSP groups was compared between the baseline and challenge scans. No difference between GSP and PSP in voxel intensity was observed in the PFC.

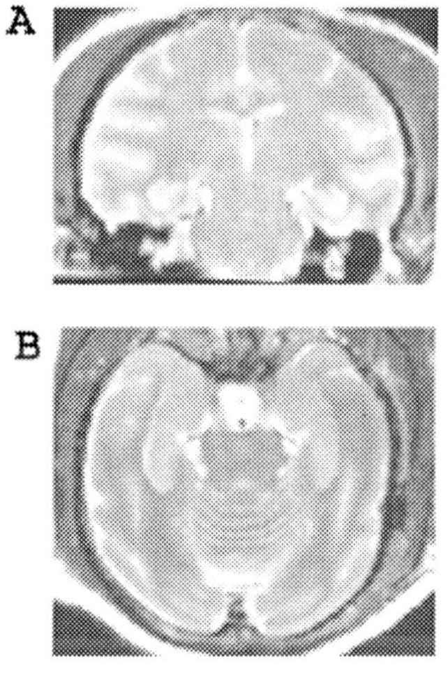

Figure 4. Representative structural MRI images from the hippocampus in A. coronal and B.horizontal views. Like the PFC, in the hippocampus, regions of interest identified the hippocampus, using a functional overlay. Voxel intensity between the baseline and scopolamine challenge scans were compared for each animal. GSP monkeys demonstrated a greater difference in voxel intensity between the baseline and scopolamine challenge scan compared to PSP monkeys.

Regions of Interest (ROI) were drawn for the PFC and hippocampus based on the anatomical scan using key neuroanatomical landmarks from anatomical scans.

The ROI was drawn for each animal and was the same ROI for the baseline and scopolamine scan, both of which were co-registered to the anatomical scans. In the phMRI, control animals administered isotonic saline did not show a significant change in BOLD signal in the PFC or hippocampus. In the hippocampus, but not the PFC, the baseline BOLD signal was significantly lower in the GSP compared to the PSP group. Following scopolamine administration, the BOLD signal change was significantly greater in the GSP than the PSP group in the hippocampus but not the PFC. The phMRI results suggest that there is a different neuronal connectivity of the muscarinic receptor system in the hippocampus compared to the PFC, based on the baseline BOLD signal, and that the muscarinic receptor system in the hippocampus was more functional in the GSP than the PSP group.

Pharmacological intervention and anesthesia are important variables to consider with phMRI in animals. We describe an experiment with scopolamine, a nonspecific muscarinic receptor antagonist.

The half life of scopolamine is 17 min in plasma [43]. However, it crosses the blood brain barrier quickly [44] and the effects in the CNS are present for up to 8 hours [45]. Peripherally administered scopolamine binds to muscarinic receptors within the brain [43, 46] and alters EEG activity in healthy, resting humans [47]. In order to anesthetize the animals, we used isoflurane and were able to measure significant BOLD signal changes.

This is consistent with other studies, demonstrating that the brain responds to stimuli during anesthesia, even at high doses [48, 49] and that isoflurane does not fully suppress EEG activity [50-52].

Experimental fMRI studies demonstrate that isoflurane anesthesia between the range of 0.9% and 1.25% preserves a measurable BOLD signal [53] level in the anesthetized monkey brain [53-55]. Isoflurane was appropriate for this phMRI study because its effects are not mediated through the cholinergic system nor does it suppress acetylcholine activity or receptor function [56, 57].

Receptor Binding Assays

To validate the phMRI results and confirm our interpretation of the results, we performed post-mortem tissue analysis on the GSP and PSP tissue

(Figure 5). For this particular data set, age-related changes in muscarinic receptor function were assessed in rhesus macaques in brain areas important for spatial learning and memory, the PFC, hippocampus, and temporal cortex. The age ranges for the monkeys were adult (5-9 years), middle-aged (11-16 years), and aged (22-30 years) male and female rhesus macaques (*Macaca mulatta*). Age-related changes were observed in the muscarinic type 1 (M_1) and type 2 (M_2) receptors in the PFC, hippocampus (Figure 6. A.) and temporal cortex. Maximum binding of M_1 receptor decreased with age, however there was no change in M_2 receptors. M_1 receptor binding affinity did not significantly change with age, yet the M_2 receptor binding affinity significantly decreased in the PFC and hippocampus [41]. Therefore, the muscarinic receptor system is differentially affected by age.

The data from the age-related changes in muscarinic binding suggest that the changes follow differing timelines in different brain regions and are not universal. Other papers have reported an age-related change in M_1 receptors in the elderly, yet these studies are slightly different from our results. The potential discrepancy in these studies is the age range used within each study. In the study described here, we used adult, middle-aged and aged animals, and our aged animals were 22-30 years. In contrast, previous groups have only used animals that were only 20 years old [24] or did not include middle-aged animals [25]. Therefore, with the increased age range, we identified more subtle age-related changes, not only from adult to middle-age, but also those that occur in the oldest-old. The middle-age and oldest-old age groups are important groups as they appear to be more reflective of the current aging human population and increase the translational properties of the study. The age-related changes in the acetylcholine system have been well documented in animal models as well as humans [13, 58] and are suggested to have an integral role in some of the neuropathology associated with Alzheimer's disease [29, 59-61]. The increase in M_1 receptors in the aged population could possibly be part of a compensatory mechanism in response to age-related decreases in acetylcholine [58, 62]. M_1, and to a lesser extent M_2, are highly expressed in areas of the brain associated with cognitive function and are hypothesized to have a role in cognitive function. Therefore, it is conceivable that elderly cognitively tested animals have a different pattern of M_1 and M_2 expression, dependent on their cognitive performance.

To determine if there were individual differences in the M_1 and M_2 maximum receptor binding and receptor affinity potentially reflective of spatial performance, we performed M_1 and M_2 receptor binding assays on PFC and hippocampus tissue from the cognitively tested monkeys. Receptor binding assays on the cognitively tested monkeys revealed a significantly higher maximum binding of M_1 in the hippocampus of GSP compared to PSP (Figure 6.B.). However, no difference in M_1 maximum binding was observed between GSP and PSP in the PFC. M_1 receptor affinity in the hippocampus was significantly higher in the GSP compared to the PSP group, yet no difference was observed between the groups in the PFC. Contrary to the M_1 difference, no difference in M_2 maximum binding or receptor affinity was observed in the PFC (Figure 7. A.). The receptor binding data support our hypothesis that the scopolamine phMRI is reflective of the neurobiology.

Comparing the results from the phMRI and receptor binding results, there is a pattern of increased BOLD signal changes, higher M_1 maximum binding, and stronger M_1 affinity, in the hippocampus but not the PFC.

Figure 5. Receptor binding assay set-up. Receptor binding technique uses a cell harvester, pictured here. Radioligands are mixed with tissue and incubated for 1 hour. The mixture is then separated using a cell harvester and radioligands bound to receptors are collected on filter paper. Scintillation fluid is added to the filter paper and the amount of radioligand present is counted in a Beta-counter to determine amount of bound radioligand present in the sample.

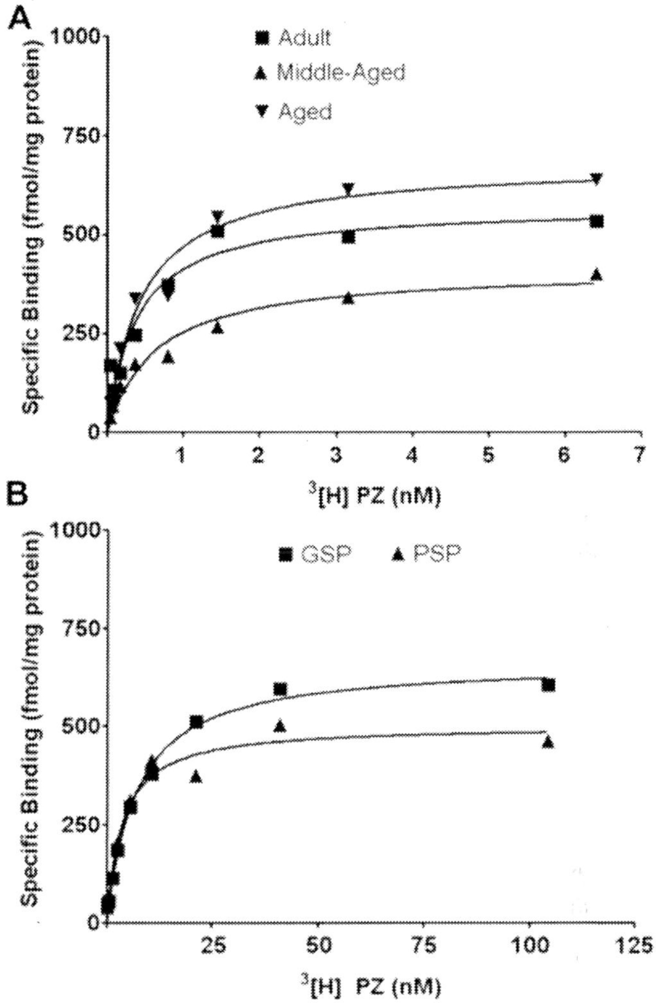

Figure 6. Receptor binding curves for muscarinic type 1 (M_1) receptor in the hippocampus. The x-axis is the number x1000 for both graphs. A. Receptor binding curves of adult (n = 5), middle-aged (n = 5), and aged monkeys (n = 5). B. Receptor binding curves of old female monkeys who were classified as GSP (n = 6) or PSP (n = 6). Age related changes are observed in the M_1 receptor binding number but not affinity. The GSP group had significantly higher receptor number and receptor affinity compared to the PSP group.

This pattern suggests that the activity of the neurons in the hippocampus of the cognitively tested female animals has a greater reliance on muscarinic receptors than the neurons of the PFC. Alternatively, it is possible the difference observed between the hippocampus and PFC is a result of experience with a hippocampus dependent task.

Following scopolamine challenge, there was a significantly greater increase in BOLD signal in the hippocampus of the GSP than PSP group, which appears to be related to cognitive performance and not age because no age differences were observed between the groups [41]. A similar pattern of the increase in BOLD signal change was observed in the PFC but it did not reach statistical significance.

A decrease in cognitive performance is observed following scopolamine injection [40, 63, 64] creating a seemingly counterintuitive increase in BOLD signal. However, scopolamine is an antagonist and antagonism of the muscarinic receptor system could initiate downstream effects on neurotransmitters, including acetylcholine, to be released locally to compensate [65, 66], thereby increasing the BOLD signal (Figure 7.B).

The neurotransmitter system could in turn initiate the recruitment of other neuronal populations, increasing neuronal activity, which would be measured as an increase in BOLD signal.

Scopolamine also targets other muscarinic receptors, including presynaptic muscarinic receptors, and blockade of those receptors can initiate the local release of acetylcholine, which is measured within the brain with microdialysis [65]. Thus it is feasible that a more specific antagonist for the M_1 receptor might have an alternate BOLD effect than scopolamine.

Alternatively, scopolamine injection could saturate muscarinic receptors in the PSP hippocampus, but was not a sufficient dose to saturate the muscarinic receptors in the GSP hippocampus; this might account for the difference between these two groups.

Based on these data, our hypothesis is that the higher number of muscarinic receptors in the GSP than PSP group creates a greater compensatory effect, resulting in a greater increase in BOLD signal following a scopolamine challenge.

Therefore, the BOLD signal following scopolamine might indicate the plasticity of the brain to compensate for a temporary blockade of the muscarinic receptor system and the brains showing more plasticity might be those which demonstrate better cognitive performance. The receptor binding assays are consistent with this hypothesis and corroborate the phMRI findings. Moreover, the receptor binding assays appear to indicate that the M_1 receptor might have a greater role in the scopolamine blockade than M_2 within the PFC and the hippocampus. In the hippocampus, the number of M_1 receptors and the receptor binding affinity is higher in the GSP than PSP group.

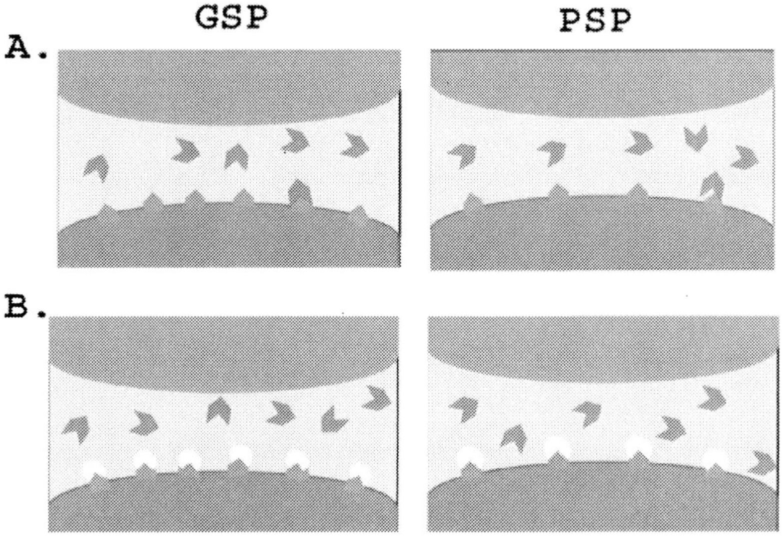

Figure 7. Schematic drawing of the hypothesized action of muscarinic receptor function in the hippocampus, based on phMRI and receptor binding assays, in good spatial performers (GSP) compared to poor spatial performers (PSP). A. In the phMRI, GSP had a greater BOLD response to scopolamine compared to PSP. At the synapse (pre-synaptic terminals in orange; post-synaptic neuron in blue), scopolamine (yellow) binds to muscarinic type 1 (M_1) receptors (green), increasing the BOLD response, potentially through increasing local acetylcholine (red). B. Receptor binding assays corroborated the phMRI indicating that the GSP had a significantly greater number of M_1 (green) which had a greater affinity for the ligand acetylcholine (red) compared to PSP.

Thus, a greater increase in BOLD signal in the hippocampus is associated with a greater number and binding affinity of M_1 receptors.

The M_2 receptor number and affinity was not different between the two groups, suggesting that M_2 have a minor role in spatial learning and memory [23]. In the PFC, there was more variability in the baseline BOLD signal as well as the percent change of BOLD signal compared to the hippocampus. As the groups were defined by performance in the spatial maze, this anatomical difference might be related to the fact that spatial learning and memory heavily relies on the hippocampus [67-70].

Interestingly, there was lower baseline BOLD signal in the hippocampus of the GSP than the PSP group. The lower baseline BOLD signal could have enabled a greater response measured by scopolamine injection in the GSP solely due to starting at a lower level. Conversely, the lower baseline BOLD signal in the hippocampus of the GSP group might indicate a different neuronal connectivity [71], observed when functional connectivity MRI studies are performed [72, 73], which could contribute to their enhanced cognitive performance in the spatial foodport maze [32]. Circadian activity was correlated with performance in the spatial food port maze [32].

Animals that had better sleep quality indices performed better on the spatial maze (Table 1). GSP animals had significantly higher overall activity, day activity, and Light:Dark activity ratios than PSP animals. Comparing the circadian activity data with the phMRI and receptor binding results, we found significant correlations between the percentage change in BOLD signal and Day Activity (Table 2).

Day Activity also correlated with M_1 maximum binding and receptor affinity in the hippocampus. Both correlations suggest that more Day Activity is observed in animals with greater number and higher affinity. These data support an interconnected relationship between hippocampal M_1 receptor function, spatial learning and memory, and circadian activity. Therefore, it is possible that the negative circadian effect on cognition is due to a negative circadian effect on muscarinic receptor function, in turn resulting in decreased cognitive function.

Table 2. Correlations between day time activity and percentage change in BOLD following a scopolamine challenge, M_1 maximum receptor binding and M_1 receptor affinity

Correlation	r value
Day Activity-% Change in BOLD	0.70**
Day Activity-M1 Bmax	-0.54*
Day Activity-M1 1/Kd	0.65*

*$P < 0.05$, **$P < 0.01$. A total of 12 animals were used for each correlation.

Conclusion

Collectively, the data from these studies demonstrate that 1) freely moving monkeys can learn a spatial maze, 2) spatial maze performance is associated with circadian activity in elderly female rhesus monkeys, 3) M_1 receptor function is associated with spatial performance and circadian activity, and 4) scopolamine phMRI can be used for assessment of muscarinic receptor function in the context of cognitive performance and circadian activity. The studies presented here demonstrate that grouping based on performance is valuable in identifying neurobiological differences between the groups.

As behavioral and cognitive performance might be reflective of the assessed neurobiological measures, using behavioral categorizations can yield valuable data. For the presented data, it is possible that various treatments might not yield circadian consequences but could negatively affect behavioral or cognitive functions.

It is also possible that a treatment could negatively affect M_1 receptor function and, in turn, negatively impact spatial learning and memory, yet it may not alter circadian activity due to compensatory mechanisms. Increased efforts are warranted to use behavioral categorization for translational research of neuropsychological and neurological disorders in humans to identify their underlying neurobiological consequences.

Acknowledgments

This work was supported by NIH grants RR-000163, AG-023477, AG-029612, AG-019100, AG-024978 (OHSU Roybal Center), AG-026472, and AG-027697 and an OHSU Tartar Fellowship.

References

[1] Haley, G.E., et al., Age-related decreases in SYN levels associated with increases in MAP-2, apoE, and GFAP levels in the rhesus macaque prefrontal cortex and hippocampus. *Age* (Dordr), 2010. 32(3): p. 283-96.

[2] Haley, G.E., et al., Effects of epsilon4 on Object Recognition in the Non-Demented Elderly. *Curr. Aging Sci.*, 2010. 3(2): p. 127-137.

[3] Haley, G.E. and F.W. Flynn, Agonist and hypertonic saline-induced trafficking of the NK3-receptors on vasopressin neurons within the paraventricular nucleus of the hypothalamus. *Am. J. Physiol. Regul. Integr. Comp. Physiol.*, 2006. 290(5): p. R1242-50.

[4] Haley, G.E. and F.W. Flynn, Blockade of NK3R signaling in the PVN decreases vasopressin and oxytocin release and c-Fos expression in the magnocellular neurons in response to hypotension. *Am. J. Physiol. Regul. Integr. Comp. Physiol.*, 2008. 295(4): p. R1158-67.

[5] Thathiah, A. and B. De Strooper, G protein-coupled receptors, cholinergic dysfunction, and Abeta toxicity in Alzheimer's disease. *Sci. Signal*, 2009. 2(93): p. re8.
[6] Nathanson, N.M., Synthesis, trafficking, and localization of muscarinic acetylcholine receptors. *Pharmacology and Therapeutics*, 2008. 119(1): p. 33-43.
[7] Wisman, L.A., et al., Functional convergence of dopaminergic and cholinergic input is critical for hippocampus-dependent working memory. *J. Neurosci.*, 2008. 28(31): p. 7797-807.
[8] Gage, F.H., et al., Experimental approaches to age-related cognitive impairments. *Neurobiol. Aging*, 1988. 9(5-6): p. 645-655.
[9] Fredrickson, A., et al., The use of effect sizes to characterize the nature of cognitive change in psychopharmacological studies: an example with scopolamine. *Human Psychopharmacologia*, 2008. 23(5): p. 425-436.
[10] Thomas, E., et al., Specific impairments in visuospatial working and short-term memory following low-dose scopolamine challenge in healthy older adults. *Neuropsychologia*, 2008. 46(10): p. 2476-2484.
[11] Elrod, K. and J.J. Buccafusco, An evaluation of the mechanism of scopolamine-induced impairment in two passive avoidance protocols. *Pharmacol. Biochem. Behav.*, 1988. 29(1): p. 15-21.
[12] Biggan, S.L., J.L. Ingles, and R.J. Beninger, Scopolamine differentially affects memory of 8- and 16-month-old rats in the double Y-maze. *Neurobiol. Aging*, 1996. 17(1): p. 25-30.
[13] Araujo, J.A., C.M. Studzinski, and N.W. Milgram, Further evidence for the cholinergic hypothesis of aging and dementia from the canine model of aging. *Prog. Neuropsychopharmacol. Biol. Psychiatry*, 2005. 29(3): p. 411-22.
[14] Savage, U.C., et al., Effects of scopolamine on learning and memory in monkeys. *Psychopharmacology* (Berl), 1996. 123(1): p. 9-14.
[15] Taffe, M.A., M.R. Weed, and L.H. Gold, Scopolamine alters rhesus monkey performance on a novel neuropsychological test battery. *Brain Res. Cogn. Brain Res.*, 1999. 8(3): p. 203-12.
[16] Rosier, A., L. Cornette, and G.A. Orban, Scopolamine-induced impairment of delayed recognition of abstract visual shapes. *Neuropsychobiology*, 1998. 37(2): p. 98-103.
[17] Flynn, D.D., et al., Differential alterations in muscarinic receptor subtypes in Alzheimer's disease: implications for cholinergic-based therapies. *Life Sci.*, 1995. 56(11-12): p. 869-76.

[18] Flynn, D.D., et al., Differential regulation of molecular subtypes of muscarinic receptors in Alzheimer's disease. *J. Neurochem.*, 1995. 64(4): p. 1888-91.

[19] Tamminga, C.A., The neurobiology of cognition in schizophrenia. *J. Clin. Psychiatry*, 2006. 67(9): p. e11.

[20] Rouse, S.T., et al., Localization of M(2) muscarinic acetylcholine receptor protein in cholinergic and non-cholinergic terminals in rat hippocampus. *Neuroscience Lett.*, 2000. 284(3): p. 182-6.

[21] Jagoda, E.M., et al., Regional brain uptake of the muscarinic ligand, [18F]FP-TZTP, is greatly decreased in M2 receptor knockout mice but not in M1, M3 and M4 receptor knockout mice. *Neuropharmacology*, 2003. 44(5): p. 653-61.

[22] Gautam, D., et al., M1-M3 muscarinic acetylcholine receptor-deficient mice: novel phenotypes. *J. Mol. Neurosci.*, 2006. 30(1-2): p. 157-60.

[23] Matsui, M., et al., Functional analysis of muscarinic acetylcholine receptors using knockout mice. *Life Sci.*, 2004. 75(25): p. 2971-81.

[24] Wagster, M.V., et al., Laminar Organization and Age-Related Loss of Cholinergic Receptors in Temporal Neocortex of Rhesus Monkey. *J. Neurosci.*, 1990. 10(9): p. 2879-2885.

[25] Vannucchi, M.G. and P.S. Goldman-Rakic, Age-dependent decrease in the affinity of muscarinic M1 receptors in neocortex of rhesus monkeys. *Proc. Natl. Acad. Sci. USA*, 1991. 88(24): p. 11475-9.

[26] Ma, L., et al., Selective activation of the M1 muscarinic acetylcholine receptor achieved by allosteric potentiation. *Proc. Natl. Acad. Sci.U SA*, 2009. 106(37): p. 15950-5.

[27] Fisher, A., et al., M1 muscarinic agonists can modulate some of the hallmarks in Alzheimer's disease: implications in future therapy. *J. Mol. Neurosci.*, 2003. 20(3): p. 349-56.

[28] Van Kampen, J.M. and C.B. Eckman, Agonist-induced restoration of hippocampal neurogenesis and cognitive improvement in a model of cholinergic denervation. *Neuropharmacology*, 2009. In Press.

[29] Fisher, A., Cholinergic treatments with emphasis on m1 muscarinic agonists as potential disease-modifying agents for Alzheimer's disease. *Neurotherapeutics*, 2008. 5(3): p. 433-42.

[30] Haimov, I., E. Hanuka, and Y. Horowitz, Chronic insomnia and cognitive functioning among older adults. *Behav. Sleep Med.*, 2008. 6(1): p. 32-54.

[31] Morin, C.M., F.C. Blais, and V. Mimeault, Sleep disturbances in late life, in *Handbook of clinical geropsychology*, H. M. and V.B. Van Hasselt, Editors. 1998, Plenum Press: New York. p. 273-299.
[32] Haley, G.E., et al., Circadian activity associated with spatial learning and memory in aging rhesus monkeys. *Exper. Neurol.*, 2009. 217(1): p. 55-62.
[33] Gillette, M.U., et al., Role of the M1 receptor in regulating circadian rhythms. *Life Sci.*, 2001. 68(22-23): p. 2467-72.
[34] Liu, C. and M.U. Gillette, Cholinergic regulation of the suprachiasmatic nucleus circadian rhythm via a muscarinic mechanism at night. *J. Neurosci.*, 1996. 16(2): p. 744-51.
[35] Wise, R.G. and I. Tracey, The role of fMRI in drug discovery. *J Magn. Reson. Imaging*, 2006. 23(6): p. 862-76.
[36] Knutson, B. and S.E. Gibbs, Linking nucleus accumbens dopamine and blood oxygenation. *Psychopharmacology* (Berl), 2007. 191(3): p. 813-22.
[37] Zhang, Z., et al., Assessing nigrostriatal dysfunctions by pharmacological MRI in parkinsonian rhesus macaques. *Neuroimage*, 2006. 33(2): p. 636-43.
[38] Luan, L., et al., Pharmacological MRI (phMRI) monitoring of treatment in hemiparkinsonian rhesus monkeys. *Cell Transplant*, 2008. 17(4): p. 417-25.
[39] Wink, A.M., et al., Age and cholinergic effects on hemodynamics and functional coherence of human hippocampus. *Neurobiol. Aging*, 2006. 27: p. 1395-1404.
[40] Thiel, C.M., Cholinergic modulation of learning and memory in the human brain as detected with functional neuroimaging. *Neurobiol. Learning Memory*, 2003. 80: p. 234-244.
[41] Haley, G.E., et al., *Hippocampal M1 Receptor Function Associated with Spatial Learning and Memory in Aged Female Rhesus Macaques AGE*, 2010. In Press.
[42] Ali-Melkkila, T., J. Kanto, and E. Iisalo, Pharmacokinetics and related pharmacodynamics of anticholinergic drugs. *Acta Anaesthesiol. Scand.*, 1993. 37(7): p. 633-642.
[43] Frey, K.A., et al., Quantitative in vivo receptor binding: I. Theory and application to the muscarinic cholinergic receptor. *J. Neurosci.*, 1985. 5(2): p. 421-428.
[44] Frey, K.A., B. Ciliax, and B.W. Agranoff, Quantitative in vivo receptor binding. IV: Detection of muscarinic receptor down-regualtion by

equilibrium and by tracer kinetic methods. *Neurochem. Res.*, 1991. 16(9): p. 1017-1023.

[45] Ali-Melkkila, T., J. Kanto, and E. Iisalo, Pharmacokineticss and related pharmacodynapics of anticholinergic drugs. *Acta Anaesthesiol. Scand.* 1993. 37(7): p. 633-642.

[46] Wamsley, J.K., et al., Muscarinic cholinergic receptors: autoradiographic localization of high and low affinity agonist binding sites. *Brain Res.*, 1980. 200: p. 1-12.

[47] Ebert, U., et al., Pharmacokinetic-pharmacodynapic modeling of the electroencephalogram effects of scopolamine in healthy volunteers. *J. Clin. Pharmacol.*, 2001. 41: p. 51-60.

[48] Porkkala, T., et al., Median nerve somatosenory evoked potentials during isoflurane anaesthesia. *Can. J. Anaesthesia*, 1997. 44: p. 963-968.

[49] Hartikainen, K. and M. Rorarius, Cortical responses to auditory stiuli during isoflurane burst suppression. *Anaesthesia*, 1999. 54: p. 210-214.

[50] Tsushima, K., et al., Supressive actions of volatile anaesthetics on the response capability in cats. *Can. J. Anaesthesia*, 1998. 45: p. 240-245.

[51] Ogawa, T., et al., The divergent actions of volatile anaesthetics on background neuronal activity and reactive capability in the central nervous system in cats. *Can. J. Anaesthesi*a, 1992. 39: p. 862-872.

[52] Murrell, J.C., D. Waters, and C.B. Johnson, Comparative effects of halothane, isoflurane, sevoflurane, and desflurane on the electroencephalogram of the rat. *Laboratory Animals*, 2008. 42: p. 161-170.

[53] Vincent, J.L., et al., Intrinsic functional architecture in the anaesthetized monkey brain. *Nature*, 2007. 447: p. 83-86.

[54] Logothetis, N.K., et al., Functional imaging of the monkey brain. *Nature Neurosci.*, 1999. 2(6): p. 555-562.

[55] Small, S.A., et al., Imaging correlates of brain function in monkeys and rats isolates a hippocampal subregion differentially vulnerable to aging. *Proc. Natl. Acad. Sci. USA*, 2004. 101(18): p. 7181-7186.

[56] Eger, E.I., et al., Acetylcholine receptors do not mediate the immobilization prodcued by inhaled anesthetics. *Anaesthesia and Analgesia*, 2002. 94: p. 1500-1504.

[57] Nakayama, T., et al., Differential effects of volatile anesthetics on m_3 muscarinic receptor coupling to the $G\square_q$ heterotrimeric G protein. *Anesthesiology*, 2006. 105: p. 313-324.

[58] Sarter, M. and J. Turchi, Age- and dementia-associated impairments in divided attention: psychological constructs, animal models, and

underlying neuronal mechanisms. *Dement. Geriatr. Cogn. Disord.*, 2002. 13(1): p. 46-58.

[59] Shiozaki, K., et al., Distribution of m1 muscarinic acetylcholine receptors in the hippocampus of patients with Alzheimer's disease and dementia with Lewy bodies-an immunohistochemical study. *J. Neurol. Sci.*, 2001. 193(1): p. 23-8.

[60] Pakaski, M. and J. Kalman, Interactions between the amyloid and cholinergic mechanisms in Alzheimer's disease. *Neurochem. Int.*, 2008. 53(5): p. 103-11.

[61] Lai, M.K., et al., Selective effects of the APOE epsilon4 allele on presynaptic cholinergic markers in the neocortex of Alzheimer's disease. *Neurobiol. Dis.*, 2006. 22(3): p. 555-61.

[62] Costall, B., et al., Biochemical models for cognition enhancers. *Pharmacopsychiatry*, 1990. 23 Suppl 2: p. 85-8; discussion 89.

[63] Klinkenberg, I. and A. Blokland, The validity of scopolamine as a pharmacological model for cognitive impairment: A review of animal behavioral studies. *Neurosci. Biobehav. Rev.*, 2010. In press.

[64] Ebert, U. and W. Kirch, Scopolamine model of dementia: electroencephalogram findings and cognitive performance. *Eur. J. Clin. Invest.*, 1998. 28(11): p. 944-9.

[65] Day, J., G. Damsma, and H.C. Fibiger, Cholinergic activity in the rat hippocampus, cortex and striatum correlates with locomotor activity: an in vivo microdialysis study. *Pharmacol. Biochem. Behav.*, 1991. 38(4): p. 723-9.

[66] Scali, C., et al., Peripherally injected scopolamine differentially modulates acetylcholine release in vivo in the young and aged rats. *Neurosci. Lett.*, 1995. 197(3): p. 171-4.

[67] Raber, J., et al., Radiation-induced cognitive impairments are associated with changes in indicators of hippocampal neurogenesis. *Radiat. Res.*, 2004. 162: p. 39-47.

[68] Cho, Y.H., E. Friedman, and A.J. Silva, Ibotenate lesion of the hippocampus impair spatial learning but not contextual fear conditioning. *Behav. Brain Res.*, 1999. 98: p. 77-87.

[69] Banta Lavenex, P. and P. Lavenex, Spatial memory and the monkey hippocampus: not all space is created equal. *Hippocampus*, 2009. 19(1): p. 8-19.

[70] Lavenex, P.B., D.G. Amaral, and P. Lavenex, Hippocampal lesion prevents spatial relational learning in adult macaque monkeys. *J. Neurosci.*, 2006. 26(17): p. 4546-58.

[71] Fransson, P., Spontaneous low-frequency BOLD signal fluctuations: An fMRI investigation of the resting-state default mode of brain function hypothesis. *Human Brain Mapp*ing, 2004. 26(1): p. 15-29.

[72] Qiu, A., et al., Hippocampal-cortical structural connectivity disruptions in schizophrenia: An integrated perspective from hippocampal shape, cortical thickness, and integrity of white matter bundles. *Neuroimage*, 2010. In Press.

[73] Pereira, F.R., et al., Asymmetrical hippocampal connectivity in mesial temporal lobe epilepsy: evidence from resting state fMRI. *BMC Neurosci.*, 2010. 11: p. 66.

In: Primates
Editors: E. F. Hughes and M. E. Hill
ISBN: 978-1-62081-498-7

Chapter V

Comparing Methods for Assessing Learning and Cognition in Primates*

Michael J. Beran, Theodore A. Evans and David A. Washburn
Georgia State University, GA, US

Abstract

One of the dominant research areas in behavioral primatology pertains to assessments of the cognitive capacities of various nonhuman primate species. Experiments in this area have shown that primates have many skills that approximate those found in humans. These tests include both manual and computerized assessments, and often the format in which tests are presented can impact the learning and behavioral responding that are produced by subjects. In this chapter, we outline two case studies that compare different methodologies and how those methodologies might contribute to differential learning and cognitive performance. In the first, rhesus monkeys responded during a learning task using either joysticks or touch-screens. The task assessed whether

* A version of this chapter also appears in Primatology: Theories, Methods and Research, edited by Emil Potocki and Juliusz Krasiński, published by Nova Science Publishers, Inc. It was submitted for appropriate modifications in an effort to encourage wider dissemination of research.

the monkeys responded according to stimulus-response associative processes or by using more relational rules about the task. Performance was similar using both input methods and suggested a more associative response pattern. In the second case, chimpanzees and monkeys learned to sequence stimuli using either a computerized task or through selection of three-dimensional stimuli presented in a manual test. Here, the manual test produced faster acquisition and better performance than the computer task, indicating that in some contexts a hands-on, real world format may produce improved cognitive performance.

Introduction

In the past few decades, the resurgence of interest in cognitive processes (the so-called Cognitive Revolution within psychology) has crossed the species boundary to include tests with various nonhuman animal species (e.g., Tomasello and Call, 1997; Wasserman and Zentall, 2006). No group of animals has been more widely studied in this domain than the nonhuman primates (e.g., Maestripieri, 2003). Tests with these species include prosimians, New World monkeys, Old World monkeys, and apes, and they include investigations into processes including perception (e.g., Beran, 2006; Dufour, Pascalis, and Petit, 2006; Fujita, and Giersch, 2005; Hopkins and Washburn, 2002; Morris and Hopkins, 1993; Tomonaga, 1998; Zivotofsky, Goldberg, and Powell, 2005), categorization (e.g., Bovet and Washburn, 2003; Santos, Hauser, and Spelke, 2001; Tanaka, 2001; Smith, Minda, and Washburn, 2004; Vonk and MacDonald, 2004), long-term memory (e.g, Beran, 2004, Beran, Pate, Richardson, and Rumbaugh, 1999); episodic memory (e.g., Hampton, Hampstead, and Murray, 2005; Menzel, 1999; Schwartz, Hoffman, and Evans, 2005), numerical processing (e.g., Beran, 2001; Beran and Rumbaugh, 2001; Brannon and Terrace, 2000; Cantlon and Brannon, 2007; Jordan, MacLean, and Brannon, 2008; Nieder and Merten, 2007; Tomonaga, 2007), planning (e.g., Beran, Pate, Washburn, and Rumbaugh. 2004; Biro and Matsuzawa, 1999; Osvath and Osvath, 2008), and metacognition (Beran, Smith, Redford, and Washburn, 2006; Call and Carpenter, 2001; Hampton and Hampstead, 2006; Mulcahy and Call, 2006; Smith, Beran, Redford, and Washburn, 2006; Smith, Shields, and Washburn, 2003; Suda-King, 2007). A wide variety of test paradigms have been developed, including many that rely on computer-based apparatus and software.

Much of the early psychological testing of captive nonhuman primates and other animals primarily relied on manual devices such as the Puzzle Box (Thorndike, 1898) and the WGTA (Harlow and Bromer, 1938). Occasional electronic devices were used for such tests (e.g., Ferster, 1964), but the widespread use of computerized testing of nonhuman primates has occurred only in the past 20 years. Computerized testing received a boost as more efficient and user-friendly data collection software emerged (e.g., Cohen, MacWhinney, Flatt, and Provost, 1993; Stahl, 2006; Washburn, 1990). These advances decreased the time it took to complete experiments by allowing faster creation, implementation, and data collection in comparison to the use of manual devices (Richardson, Washburn, Hopkins, Savage-Rumbaugh, and Rumbaugh, 1990; Washburn and Rumbaugh, 1992). However, there remains little research directly comparing performance on computerized versus manual tests of the same cognitive and learning processes. In addition, even with the adoption of the computer as a testing method with nonhuman primates, there remain questions about the best response input methods and the potential differences in performance that can occur as a function of that input method.

Comparing Manual and Computerized Test Paradigms

Perhaps the most famous apparatus in behavioral primatology is Harlow's Wisconsin General Test Apparatus (WGTA). This apparatus (and the many variations of it used since its introduction) allows an experimenter to present stimuli to an animal and record its responses without the possibility of providing inadvertent cues. Harlow (1949) and others have used this apparatus to present monkeys with 2-choice discrimination problems. Despite its innovation, however, the WGTA required an experimenter to manually present stimuli to subjects and record data. It was labor intensive, and data collection was slow. The introduction of computerized testing changed that situation. Today, many laboratories present nonhuman primates with visual tasks on computer monitors, in which individuals are rewarded, via automated feeders, for making joystick or touch-screen responses to task stimuli. The computerized method has several advantages over alternatives such as the WGTA. These include:

- The experimenter can gain complete control over the testing environment by producing stimuli and contingencies that are consistent between trials/sessions.
- Response accuracy and latency can be recorded by computer software with greater precision, leaving little room for human error.
- The experimenter does not have to be present for testing, further removing any potential cue-giving and requiring less time (on the experimenter's part) to acquire data. This is particularly important during the training of new animals, during which the computer software can adjust itself to the competence level of each subject and modify the difficulty of the training tasks.
- Given the speed with which new trials can be prepared and presented (which can be almost immediately), the data set that results is also usually much greater in size, per unit time.
- The computerized method can be easily adapted for use with different primate species, as well as some non-primate animal models.
- The task parameters can be automatically adjusted by the software to accommodate the proficiency of subject being tested.

Computerized tests of nonhuman primate cognition have offered much data in support of high level performance. In some cases, the use of computerized tests produced radically different behavioral outcomes compared to tests given with manual or other types of non-computerized tests. For example, Washburn and Rumbaugh (1991b) found that monkeys, when tested with a large number of computerized reversal learning problems, exhibited a more cognitive learning style than when tested previously with a small number of manually presented problems (Essock-Vitale, 1978). McGonigle, Chalmers, and Dickinson (2003) also reported that capuchin monkeys performed better on tests of seriation and classification when those tests were computerized compared to manual tests.

However, in other cases manual tests have produced stronger evidence of cognitive competence. Mandrell and Sackett (2008), for example, tested infant macaque monkeys on a variety of computerized tests and then compared performance to a substantial database of previous similar tests given in a standard manual format (WGTA). They found that for some tasks, performance was comparable between computerized and manual tests. However, in a rule learning task (learning set), the animals tested on the computer version performed more poorly compared to animals tested with the WGTA. These results stand in contrast to those of Washburn, Hopkins, and

Rumbaugh (1989) who reported better performance on computer versions of the learning set task compared to manual tests (although those were adult macaque monkeys, and they used joysticks rather than touch-screens, as were used with the infants). It is important to note that this type of discrepancy is not unique to tests with nonhuman primates, as human children sometimes show differences in certain tests when presented in a computer or manual format whereas in other cases, and for the same task, they show no such differences (e.g. Feldstein et al., 1999; Carlson, Moses, and Claxton, 2004; Luciana and Nelson, 2002).

Comparing Joystick and Touch-Screen Input on Computerized Tasks

Despite the advantages offered by computerized testing, once a research team decides to use a computerized task, they must still determine the appropriate input device. Typically, primates respond using either a joystick or a touch-screen. In labs in which animals are provided with a joystick, they can manipulate the joystick with their hand to control the movement of a cursor on the screen or they can simply deflect (or hold) the joystick to make different kinds of behavioral responses. In labs in which primates are provided with touch-screen monitors, they contact visual stimuli directly with their hands.

Other important factors should be considered when choosing between touch-screens and joysticks. Each input works considerably well with different primate species, but they differ in expense, additional equipment needed, and the type of software needed for testing (most of these favoring the joystick). Also, each device can produce unique methodological issues that need to be considered when designing or interpreting an experiment (see Evans et al., 2008). One of these issues is stimulus contact accuracy. When using a touch-screen, it is easier for animals to make direct responses to stimuli, and touch-screens require less fine motor coordination to select stimuli. With joysticks, more psychomotor control is needed to select stimuli, and so this should be considered in relation to the task parameters themselves (e.g., the number, size, and spatial proximity of the stimuli presented on the screen). Another difference between input devices is their ability to accurately represent and record animal response time and decision time. Touch-screens can be more accurate in this respect because of the ease with which animals can select stimuli. When joysticks are used, one must consider the possibility that response times may vary because of less-than efficient cursor control, in addition to time required to make a decision.

An important difference between these two input devices is the level of spatial contiguity between the response location and the stimulus location. The touch-screen provides perfect spatial contiguity, while the joystick does not. Some researchers have argued that differences in the level of spatial contiguity between stimuli and response loci can result in differences in performance on psychological assessments. For example, rhesus macaques (*Macaca mulatta*) often had difficulty in learning two-choice discrimination tasks when there was stimulus-response (S-R) spatial discontiguity using a manual apparatus like the WGTA (McClearn and Harlow, 1954; Meyer, Treichler, and Meyer, 1965; Murphy and Miller, 1955, 1958). Meyer, Polidora, and McConnell (1961) reported that performance of monkeys was poorer when responding to stimuli discontiguous with an activating panel in a computerized apparatus than when the response loci were contiguous with the cue stimuli.

However, rhesus monkeys have overcome spatial discontiguity when testing involved joystick-based computerized tests (Richardson et al., 1990; Rumbaugh, Richardson, Washburn, Savage-Rumbaugh, and Hopkins, 1989; Washburn et al., 1989). Beran, Rumbaugh, and Washburn (2007) reported that rhesus monkeys even overcame second-order S-R spatial discontiguity when presented with Harlow's (1949) learning set task. The monkeys used a joystick to move a cursor so as to contact computer graphic "levers" (which differed in location across experimental phases) in order to select an associated test stimulus. The levers were spatially discontiguous with the response stimuli themselves (i.e., the S+ and S- stimuli to be discriminated). Therefore, the task produced both first-order (joystick and lever) and second-order (lever and stimuli) spatial discontiguity. These performances showed that rhesus macaques do not attend simply to the region around the cursor in these computerized tests. They also attend to relevant stimulus loci even when these are discontiguous with response and reward areas, and they do so when tested using joystick methodology.

In summary, computerized testing methods offer several advantages over traditional manual apparatus methods, but this does not mean that computer tests always afford the greatest likelihood of finding positive results or the clearest evidence of some form of learning or cognition. The benefits of computer testing must be weighed against the possibility that such tests might fail to tap into important subject variables. Despite some evidence that computer tests have helped overcome previous failures of learning, it is possible that other kinds of behavioral and cognitive phenomena might not manifest as well in computerized paradigms. For example, whereas computerized tests sometimes show enhanced cognitive performance in children (e.g., Olsen, 2008), in other cases the use of computer tasks shows less cognitive sophistication in the resulting performance of children (Luciana and Nelson, 2002).

In this paper, we outline two research projects designed to assess differences in performance as a function of methodology used with primates. In the first, we directly compared performance on the same computerized learning task when presented as either a touch-screen task or a joystick task. This task, the mediational paradigm, illustrates the extent to which participants can flexibly learn discrimination rules and then modify those rules with experience in responding to pairs of stimuli. In the second project, we compared performance in an ordinal learning task (the sequencing of arbitrary stimuli on the basis of the quantitative relations to each other) when it was presented in a computerized format versus a more naturalistic, manual format. These projects will help illustrate the differences in cognitive and learning performances as a function of the methodology that is used.

Project 1. Mediational Paradigm Performance with Joysticks and Touch-Screens

Rumbaugh (1971) devised the mediational paradigm specifically for differentiating associative and rule-based interpretations for reversal learning in nonhuman animals. Performance on this task would indicate whether animals responded through stimulus-response (S-R) associative learning processes such as those posited for more general learning phenomena (Hull, 1943; Spence, 1937) or whether they learned about relations between stimuli by discovering the logical task structuring. The mediational task was based on the well-established phenomenon of learning set (Harlow, 1949), and it has been used with a variety of species (e.g., Beran et al., 2008a; Rumbaugh and Pate, 1984a, 1984b).

Here, we assessed performance in four rhesus monkeys (Obi – age 2 years, Han, age 3 years, Chewie, age 6 years, and Luke, age 6 years). Prior to testing, each animal received training with each input device (touch-screen and joystick). These training procedures have been extensively outlined elsewhere (e.g., Evans et al., 2008; Richardson et al., 1990). We will cover the basic details only. It is important to note that all monkeys were trained on both input forms before this study, so their experiences when entering this task were identical, and performance differences on the mediational paradigm could not differ as a result of different experiences with computer tasks.

When monkeys were trained to use the joystick, they progressed through a seven step procedure. First, they were shaped to simply touch, and then displace, the joystick. These steps usually occurred automatically and required little or no active training. Monkeys then were trained across several steps to move the cursor to smaller and smaller portions of the screen area. They then were trained to contact a moving target with the cursor. Finally, they learned to select stimuli in a matching-to-sample paradigm in which a visual sample was presented mid-screen and an identical stimulus had to be selected from the perimeter of the screen. A similar set of steps were taken to train touch-screen use. Monkeys first were rewarded to touch any part of the computer screen, and then were rewarded to touch sections of the screen that were successively smaller in size as training criteria were reached. They then were trained to touch a moving target, and then to touch stimuli in the matching-to-sample paradigm. At the end of training on the input modalities, two monkeys (Luke and Obi) were presented with the mediational paradigm and responded using the touch-screen. The other two monkeys (Chewie and Han) responded using a joystick. Monkeys were housed individually, and had 24-hour access to the task, so they could work whenever they chose. For two weeks the mediational paradigm was presented continuously, and the monkeys completed thousands of trials using either the joystick or touch-screen. The task was identical as presented to the two groups except that the joystick monkeys moved a cursor on the screen into contact with choice stimuli whereas the touch-screen monkeys physically touched those choice stimuli with their fingers or hands. Correct responses provided the monkeys with a food pellet reward whereas incorrect responses led to a time-out period during which the screen was blank and no responses could be made. The mediational paradigm itself consisted of a number of unique, discrete "problems" that involved the repeated presentation of pairs of stimuli. Each problem consisted of three steps:

1) In the training step, a monkey was provided with a pair of stimuli: an S+ that, when selected, would result in a food reward, and an S- that, when selected, would result in a time-out penalty during which the screen was blank. These stimuli were denoted A+ and B-. Training trials were presented until the subject reached a 90% criterion over the last 10 trials.

2) After reaching this criterion, a single trial (called the reversal trial) was presented in which the response contingencies for stimuli A and B were reversed, making A the new S- and B the new S+.
3) After the single reversal trial, stimuli were presented in 10 test trials in one of three possible conditions. In the first condition, A and B continued to be tested with reversed response contingencies (as S- and S+, respectively). In the second and third conditions, a novel stimulus, C, replaced either A or B as new S- or the New S+, respectively. At the end of these 10 test trials, a new problem was presented (with two new stimuli, arbitrarily assigned as S+ and S-), and the training phase began again.

A comparison of performance in the three conditions provides evidence for the type of learning that occurs. If an animal has learned only to approach the A stimulus during training (A+B-), the B+C- condition should produce the best performance levels because there is no need to extinguish approach responses to stimulus A because it is no longer present.

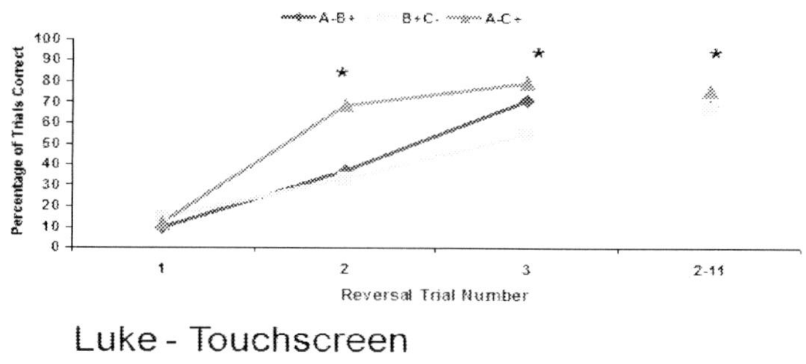

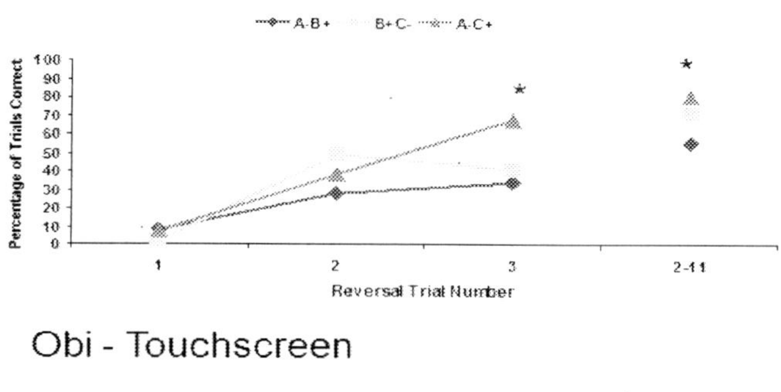

Figure 1. Performance in each condition by the two monkeys that responded using a touch-screen. The single A-B+ reversal trial that occurred between training and test trials is labeled here as Trial 1, and the 10 test trials are labeled as Trials 2-11.

If an animal has learned only to avoid the B stimulus, the A-C+ condition should produce the best performance levels because there is no need to overcome inhibition of responding to stimulus B. If the animal has learned both to avoid the B stimulus and approach the A stimulus, performance should be poorest on the control (A-B+) condition because both associative processes (approach and inhibition) remain. These are all predictions based on associative learning as the primary force behind the animals' behavior. However, a more cognitive interpretation also exists. If the monkeys learn more than inhibitory and excitatory associations formed between specific stimuli and responses, they could discern all that they needed to know from the first reversal trial. Specifically, that trial illustrates that now the B stimulus is the correct stimulus and the A stimulus is the incorrect stimulus, and the subsequent introduction of the C stimulus would not disrupt performance.

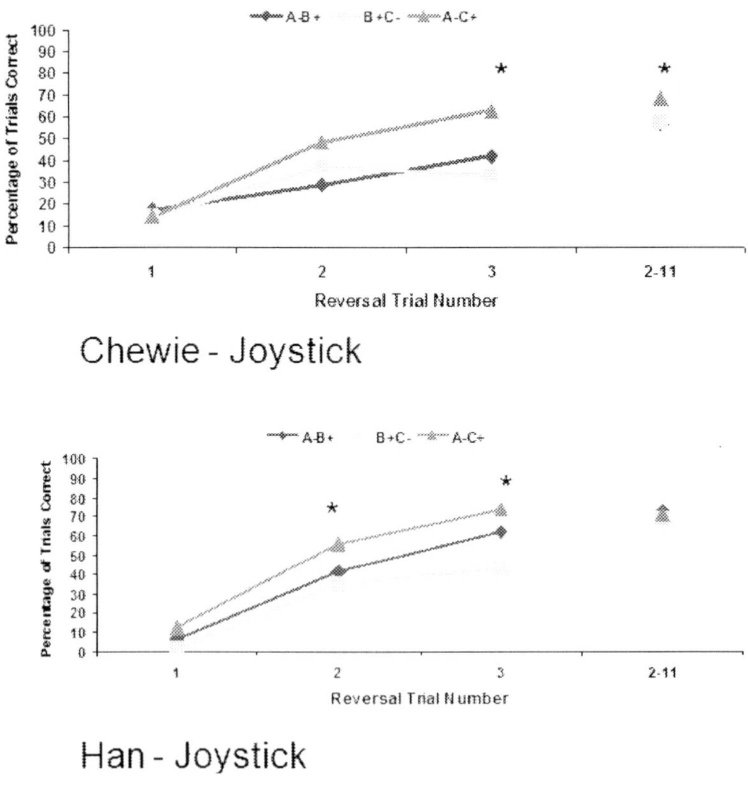

Figure 2. Performance in each condition by the two monkeys that responded using a joystick. The single A-B+ reversal trial that occurred between training and test trials is labeled here as Trial 1, and the 10 test trials are labeled as Trials 2-11.

In the B+C- condition, the organism could continue to respond based on learning that B is now the correct response. In the A-C+ condition, the organism could respond based on learning that the A stimulus is no longer the correct response, and thus the C stimulus should be selected. The A-B+ condition would allow a continuation of the win-stay, lose-shift rule that emerged from the first reversal trial (see Rumbaugh and Pate, 1984a, 1984b).

Performance for all four animals was very similar, and in line with a previous experiment conducted with joystick-using capuchin monkeys (Beran et al., 2008). The touch-screen monkeys (Figure 1) each performed significantly differently on the three conditions across all trials after the reversal trial, Luke $F(2, 305) = 4.56, p < .05$; Obi $F(2, 150) = 12.92, p < .01$. Luke also showed a significant difference in performance for Trial 3 after reversal and Obi showed a significant difference in performance for Trials 2 and 3 after the reversal trial. The joystick monkeys showed slightly more variability in performance (Figure 2). Chewie performed significantly differently on the three conditions across all trials after the reversal trial, $F(2, 206) = 30.59, p < .01$, but Han did not $(2, 228) < 1.00, p = .63$. However, Han showed a significant difference in performance for Trial 2 and Trial 3 after reversal, and Chewie showed a significant difference in performance for Trial 3 after the reversal trial. The touch-screen monkeys and joystick monkeys all appeared to learn the response contingencies of test stimuli faster in the A-C+ condition than in the A-B+ or B+C- condition. Thus, the monkeys performed the two-choice discrimination faster when the S+ was replaced with a novel stimulus, and the S- remained the same. This suggested that all of the monkeys learned to inhibit responding to the S- during training because this stimulus produced punishment (in the form of a time-out penalty). These data also suggested that the input device used for testing, and thus the level of stimulus-response spatial contiguity included in the task, did not influence learning performance. In other words it made no difference whether the response occurred at the same location as the stimulus or at a different location. Performance was qualitatively similar with both types of response input.

Project 2. Acquiring Ordinal Sequences Quickly – Manual versus Computerized Tasks

Ordinality refers to the relative position of one entity with respect to other entities in a sequence. Some research programs have trained animals to associate various symbolic stimuli with discrete quantities. For example,

Washburn and Rumbaugh (1991) trained rhesus monkeys to select among 2, 3, 4, or 5 Arabic numerals presented on a computer screen, and the monkeys received an equivalent number of food pellets to each numeral that was selected. Those monkeys selected the larger numeral when presented with pairs of numerals, and they succeeded when faced with novel arrays of 3, 4, or 5 different numerals at one time. Beran et al. (2008b) replicated these findings with capuchin monkeys and rhesus monkeys. Olthof, Iden, and Roberts (1997) also presented monkeys with two numerals and the monkeys received a number of food items equal to the numeral that they selected on each trial. Olthof and Santi (2007) used similar tests with pigeons, and they also showed that ordinal relations emerged from this pair-wise training paradigm with arbitrary stimuli.

In most studies that use this task, acquisition of the ordinal series required a large number of training trials. Washburn and Rumbaugh (1991a) reported that their monkeys needed more than 400 trials to reach high levels of performance on the ordinal task. Beran et al. (2008b) gave their monkeys 1,000 training trials before moving to the test phase. Olthof et al. (1997) reported that their monkeys required a large number of training trials, and pigeons required 45 multiple-trial training sessions to reach criterion in selecting the larger member of a pair of presented symbols (Olthof and Roberts, 2000). Other research projects that required nonhuman animals to select visible sets of non-food items (i.e., magnitudes) in ordinal sequences also reported that large numbers of training trials were required (e.g., Brannon and Terrace, 2000; Judge, Evans, and Vyas, 2005; Smith, Piel, and Candland, 2003).

One study, however, reported very rapid acquisition of an ordinal sequence of arbitrary stimuli. Beran, Beran, Harris, and Washburn (2005) reported that two chimpanzees and a rhesus macaque rapidly learned the ordinal relations between five colors of containers (plastic eggs) when all containers of a given color contained a specific number of identical food items. All three animals also performed at high levels when comparing sets of containers to sets of visible food items. In that paradigm, the animals were presented with two different colored plastic eggs, each containing some number of food items. The color of the egg indicated how many food items were inside, and so colors constituted the ordinal sequence. These primates exceeded chance levels within the very first 40-trial session, and they reached near perfect performance within only 3 or 4 such sessions (<150 trials). In addition, the chimpanzees successfully selected among 3, 4, and 5 egg arrays in the correct decreasing quantity order from the very first presentations of such trials. Therefore, very rapid acquisition of the desired ordinal sequence was established.

In the present experiment, we first determined whether the chimpanzees and the monkey in the Beran et al. (2005) study remembered the learned ordinal sequence of these egg stimuli after a 9-month delay. These data would provide evidence of long-term memory for the ordinal sequence, matching other previous reports of such long-term memory in animals (Beran, 2004; Beran et al., 2000; Burdyn, Noble, Shreves, and Thomas, 1984; Patterson and Tzeng, 1979; Reichmuth Kastak and Shusterman, 2002). Then, we tested chimpanzees and monkeys on variations of the ordinal learning task. We used the manual test with plastic eggs that physically held food, and we compared performance on that test to performance on a computerized variation in which colored squares were selected on the screen (with resulting delivery of different quantities of food items as a result of the color that was selected). This experiment allowed us to determine whether the specific format of the Beran et al. (2005) ordinal training led to this rapid learning of the sequence in comparison to other kinds of tests that have been used (namely, those presented in a computer task).

9-Month Retest with the Manual Task

In the original project (Beran et al., 2005) two chimpanzees (Sherman and Lana) and a rhesus monkey (Gale) were presented with eggs in five different colors – red, blue, green, orange, and yellow. Those five colors always represented 5, 4, 3, 2, and 1 food reward contained within the respective eggs of those colors. Training took the form of presenting four trials in each session of all possible combinations of two of those eggs (with the larger value egg equally often on each side of the presentation apparatus). To reach the criterion for mastery of the ordinal sequence, an animal had to be correct on 90% of trials within a session. All three animals exceeded chance levels on the very first session. Sherman and Gale required only three sessions to reach the criterion, and Lana required four sessions.

In the re-test, each of these primates completed a single 40-trial session just like those used during training. The animals had not seen the eggs or performed this task for more than 9 months. Lana and Gale were perfect, choosing the larger-valued egg on all 40 trials. Sherman made a single error and was 97.5% correct overall. In a second session, the animals were presented with a single egg compared to a visible quantity of food (this condition was used in the original study to demonstrate that the animals had learned something about the absolute number of food items in each egg). Sherman was correct on 92.5% of these trials. Lana was correct on 85% of these trials, and Gale was correct on 72.5% of these trials. In all cases, performance exceeded chance levels as assessed with a binomial sign test, $p < .05$.

Training Ordinal Sequences of Colored Stimuli on the Computer

In the next phase, we presented the exact same task in computerized format. Now, subjects chose between two colored squares on a computer screen, and the computer dispensed the relevant number of food items on the basis of the choice that was made. The critical difference, therefore, was the use of two-dimensional stimuli that were spatially discontiguous with the response loci of the animals (the joysticks), and food rewards were not contained within those response stimuli (so, food, too was discontiguous with the response stimuli). However, the pacing of the task was the same.

We tested Sherman, Lana, and Gale, assuming they would show some transfer of learning from the egg task to the computer task because we used the same exact five colors in the same ordinal sequence. We also included two new chimpanzees (Panzee and Mercury), and two new monkeys (Willie and Hank), all of whom had no experience with the egg task, so that we could assess how much transfer actually occurred. The criterion for mastery of the ordinal sequence was the same - 90% correct over 40 trials.

The performance of each animal is shown in Figure 3. Sherman and Lana (chimpanzees) met the training criterion, although it took them many more 40-trial blocks than with the egg stimuli. In Sherman's case, performance was high from the first few blocks, although it took some time for him to finally reach the criterion (nine blocks with the computer task compared to three blocks with the egg version of the task). Lana, however, performed at chance levels for the first seven blocks, and it took her a total of 10 blocks to reach the criterion (compared to four blocks with the egg version of the task). Gale, who had met criterion quickly with the egg task and showed perfect 9-month retention, required 29 blocks of trials to meet the criterion on the computerized version of the task. He began to exceed chance levels at the 10th block of trials, but performance did not rise to criterion quickly as it had with the egg version of the task.

Of the new animals, the monkey Hank reached the criterion after 24 blocks, and he began to exceed chance levels after approximately 10 blocks. The monkey Willie reached criterion after 88 sessions although he began to exceed chance levels after approximately 60 sessions. Neither of the new chimpanzees reached criterion, although they were not given as much opportunity as the monkeys (Panzee was discontinued after 32 blocks and Mercury was discontinued after 27 blocks, because both chimpanzees only rarely exceeded chance levels of performance within a session). These results suggested that animals with no previous experience with sequencing these five colors had a difficult time learning to organize the ordinal sequence when presented with a computer version of the task. This does not mean that such a task is impossible, as primates have learned to sequence arbitrary stimuli using this paired-comparison method before (e.g., Beran et al., 2008b; Harris, Beran and Washburn, 2007; Washburn and Rumbaugh, 1991a). It just indicates that training takes substantially longer in the computerized format used here than in the more naturalistic format with the physical items that contained the food rewards. Next, we gave the four best-performing animals (Lana, Sherman, Hank, and Gale) the same computer task again but with five new colors to be learned in a new sequence. All details were the same as in the previous experiment. Lana required 23 blocks to reach the criterion, and Sherman never reached criterion even after 23 blocks (in fact, he never exceeded chance levels of responding). Hank never reached criterion (after 115 blocks), and Gale reached criterion after 31 blocks. These data indicated clearly that Sherman, Lana, and Gale did not show the rapid acquisition that they had showed with the egg stimuli in the Beran et al. (2005) study, and this computerized task was very difficult.

Finally, we re-introduced the task with the eggs. We tested all four chimpanzees. Sherman and Lana, of course, were very familiar with the task, but Panzee and Mercury had not performed this task (and, recall, both of them rarely exceeded chance levels of responding when sequencing these colors in the computerized task). As expected, Sherman and Lana met the criterion in the first session (Sherman – 100% correct; Lana – 97.5% correct). However, Mercury and Panzee also showed very rapid acquisition of the ordinal sequence. Mercury required three blocks to reach the 90% criterion, and Panzee required six blocks. Also, both new chimpanzees exceeded chance levels of responding by the end of their first session with the egg stimuli.

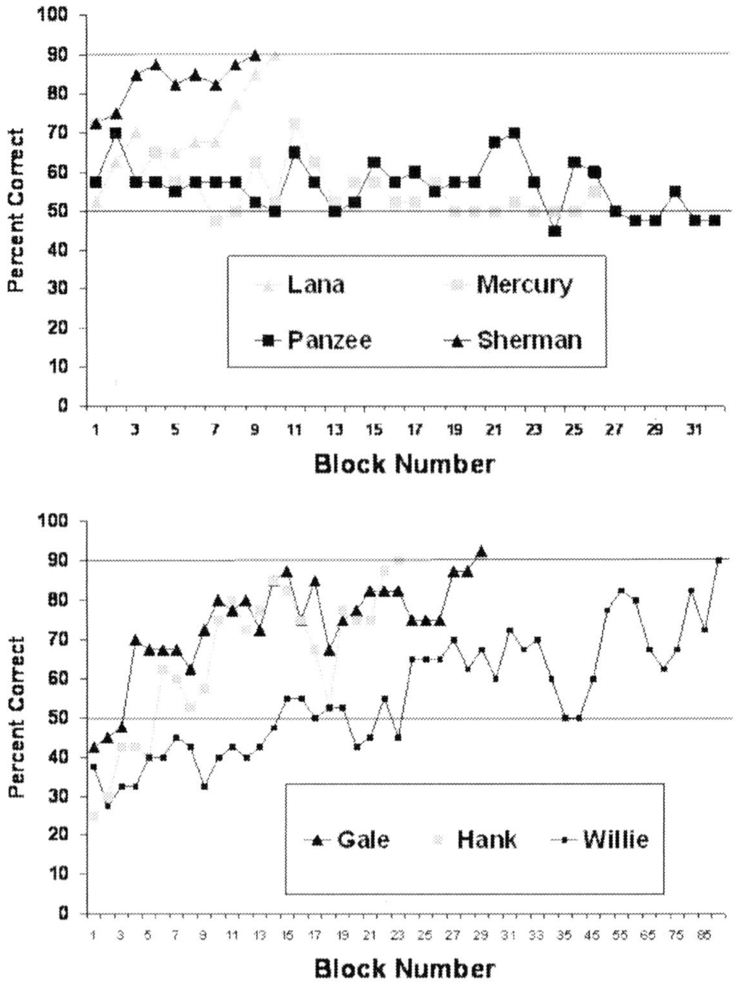

Figure 3. Performance of each animal on the computerized ordinality task. The top figure shows chimpanzee performance and the bottom figure shows monkey performance. The horizontal line at the 90% correct value indicates the criterion value, and chance is represented by the horizontal line at the 50% correct level in each graph.

Thus, we replicated the extremely rapid learning that occurs when these kinds of stimuli are used to present pair-wise training of an ordinal sequence of arbitrary stimuli. So why should this advantage for learning occur for the manual test and not the computerized test? We believe that the manual test more likely taps into the natural salience structure of the animal's environment.

Food items contained within "shells" (which the eggs were in a functional sense), and the choice of those things by direct pointing or contacting of those stimuli, more naturally mimics a foraging-like task or situation for animals such as these. This situation provides more natural salience in the association between the egg colors and the quantities that are contained within, making learning more likely on the basis of this salience (see Rumbaugh, King, Washburn, Beran, and Gould, 2007).

Therefore, this research program indicates an advantage for the non-computerized format of the ordinal learning task. These results also provide further evidence for the need for careful and insightful consideration of the best methodology and stimulus presentation format when one tries to establish a behavioral competence in a nonhuman primate species.

Conclusion

These two new research projects illustrate the care that needs to be taken when designing methods for use with nonhuman primates. In some cases, one can see flexibility as reflected in consistent performance across testing methods (as with the touch-screens and joysticks used in the mediational test). In other cases, manual versus computerized tests can produce quite different levels of performance, speed of acquisition, and retention (as in the ordinal test).

These kinds of results indicate that no single method should be relied upon solely as a means of determining whether a given individual or species does or does not have a certain behavioral trait or cognitive capacity. Rather, it is through varied and creative methodological assessments that a fuller picture of primate behavior and cognition will emerge.

Acknowledgments

We thank the many people who have contributed to this research effort over the years as colleagues and through their assistance with data collection from the primates, particularly Emily Harris, John Kelley, and Mary Beran. This research is supported by Grant HD-38051 from the National Institute of Child Health and Human Development. Correspondence can be sent to the first author at mjberan@yahoo.com.

References

Beran, M. J. (2001). Summation and numerousness judgments of sequentially presented sets of items by chimpanzees (*Pan troglodytes*). *Journal of Comparative Psychology, 115*, 181-191.

Beran, M. J. (2004). Long-term retention of the differential values of Arabic numerals by chimpanzees (*Pan troglodytes*). *Animal Cognition, 7*, 86-92.

Beran, M. J. (2006). Quantity perception by adult humans (*Homo sapiens*), chimpanzees (*Pan troglodytes*), and rhesus macaques(*Macaca mulatta*) as a function of stimulus organization. *International Journal of Comparative Psychology, 19*, 386-197.

Beran, M. J. and Beran, M. M. (2004). Chimpanzees remember the results of one-by-one addition of food items to sets over extended time periods. *Psychological Science, 15*, 94-99.

Beran, M. J. and Rumbaugh, D. M. (2001). "Constructive" enumeration by chimpanzees (*Pan troglodytes*) on a computerized task. *Animal Cognition, 4*, 81-89.

Beran, M. J., Beran, M. M., Harris, E. H. and Washburn, D. A. (2005). Ordinal judgments and summation of nonvisible sets of food items by two chimpanzees (*Pan troglodytes*) and a rhesus macaque (*Macaca mulatta*). *Journal of Experimental Psychology: Animal Behavior Processes, 31*, 351-362.

Beran, M. J., Harris, E. H., Evans, T. A., Klein, E. D., Chan, B., Flemming, T. J. and Washburn, D. A. (2008b). Ordinal judgments of symbolic stimuli by capuchin monkeys (*Cebus apella*) and rhesus monkeys (*Macaca mulatta*): The effects of differential and nondifferential reward. *Journal of Comparative Psychology, 122*, 52-61.

Beran, M. J., Klein, E. D., Evans, T. A., Chan, B., Flemming, T. J., Harris, E. H., Washburn, D. A. and Rumbaugh, D. M. (2008a). Discrimination reversal learning in capuchin monkeys (*Cebus apella*). *Psychological Record, 58*, 3-14.

Beran, M. J., Pate, J. L., Richardson, W. K. and Rumbaugh, D. M. (2000). A chimpanzee's (*Pan troglodytes*) long-term retention of lexigrams. *Animal Learning and Behavior, 28*, 201-207.

Beran, M. J., Pate, J. L., Washburn, D. A. and Rumbaugh, D. M. (2004). Sequential responding and planning in chimpanzees (*Pan troglodytes*) and rhesus macaques (*Macaca mulatta*). *Journal of Experimental Psychology: Animal Behavior Processes, 30*, 203-212.

Beran, M. J., Rumbaugh, D. M. and Washburn, D. A. (2007). Rhesus monkeys (*Macaca mulatta*) maintain learning set despite second-order stimulus-response spatial discontiguity. *Psychological Record, 57*, 9-22.

Beran, M. J., Smith, J. D., Redford, J. S. and Washburn, D. A. (2006). Rhesus macaques (*Macaca mulatta*) monitor uncertainty during numerosity judgments. *Journal of Experimental Psychology: Animal Behavior Processes, 32*, 111-119.

Biro, D. and Matsuzawa, T. (1999). Numerical ordering in a chimpanzee (*Pan troglodytes*): Planning, executing, and monitoring. *Journal of Comparative Psychology, 113*, 178-185.

Bovet, D. and Washburn, D. A. (2003). Rhesus macaques (*Macaca mulatta*) categorize unknown conspecifics according to their dominance relations. *Journal of Comparative Psychology, 117*, 400-405.

Brannon, E. M. and Terrace, H. S. (2000). Representation of the numerosities 1-9 by rhesus macaques (*Macaca mulatta*). *Journal of Experimental Psychology: Animal Behavior Processes, 26*, 31-49.

Burdyn, L. E., Noble, L. M., Shreves, L. E. and Thomas, R. K. (1984). Long-term memory for concepts by squirrel monkeys. *Physiological Psychology, 12*(2), 97-102.

Call, J. and Carpenter, M. (2001). Do apes and children know what they have seen? *Animal Cognition, 4*, 207-220.

Cantlon, J. F. and Brannon, E. M. (2007). How much does number matter to a monkey (*Macaca mulatta*)? *Journal of Experimental Psychology: Animal Behavior Processes, 33*, 32-41.

Carlson, S. M., Moses, L. J. and Claxton, L. J. (2004). Individual differences in executive functioning and theory of mind: An investigation of inhibitory control and planning ability. *Journal of Experimental Child Psychology, 87*, 299-319.

Cohen, J., MacWhinney, B., Flatt, M. and Provost, J. (1993). PsyScope: An interactive graphic system for designing and controlling experiments in the psychology laboratory using Macintosh computers. *Behavior Research Methods, Instruments, and Computers, 25*, 257-271.

Dufour, V., Pascalis, O. and Petit, O. (2006). Face processing limitations to own species in primates: A comparative study in brown capuchins, Tonkean macaques and humans. *Behavioural Processes, 73*, 107-113.

Essock-Vitale, S. M. (1978). Comparison of ape and monkey modes of problem solution. *Journal of Comparative and Physiological Psychology, 92*, 942-957.

Evans, T. A., Beran, M. J., Chan, B., Klein, E. D. and Menzel, C. R. (2008). An efficient computerized testing method for the capuchin monkey (*Cebus apella*): Adaptation of the LRC-CTS to a socially housed nonhuman primate species. *Behavior Research Methods, 40*, 590-596.

Feldstein, S. N., Keller, F. R., Portmann, R. E., Durham, R. L., Klebe, K. J. and Davis, H. P. (1999). A comparison of computerized and standard versions of the Wisconsin Card Sorting Task. *Clinical Neuropsychologist, 13*, 303-313.

Ferster, C. B. (1964). Arithmetic behavior in chimpanzees. *Scientific American, 210*, 98-106.

Fujita, K. and Giersch, A. (2005). What perceptual rules do capuchin monkeys (*Cebus apella*) follow in completing partly occluded figures? *Journal of Experimental Psychology: Animal Behavior Processes, 31*, 387-398.

Hampton, R. R. and Hampstead, B. M. (2006). Spontaneous behavior of a rhesus monkey (*Macaca mulatta*) during memory tests suggests memory awareness. *Behavioural Processes, 72*, 184-189.

Hampton, R. R., Hampstead, B. M. and Murray, E. A. (2005). Rhesus monkeys (*Macaca mulatta*) demonstrate robust memory for what and where, but not when, in an open-field test of memory. *Learning and Motivation, 36*, 245-259.

Harlow, H. F. (1949). The formation of learning sets. *Psychological Review, 56*, 51-65.

Harlow, H. F. and Bromer, J. A. (1938). A test-apparatus for monkeys. *Psychological Record, 2*, 434-436.

Harris, E. H., Beran, M. J. and Washburn, D. A. (2007). Ordinal list integration for symbolic, arbitrary, and analog stimuli by rhesus macaques (*Macaca mulatta*). *Journal of General Psychology, 134*, 183-197.

Hopkins, W. D. and Washburn, D. A. (2002). Matching visual stimuli on the basis of global and local features by chimpanzees (*Pan troglodytes*) and rhesus monkeys (*Macaca mulatta*). *Animal Cognition, 5*, 27-31.

Hull, C. L. (1943). *Principles of behavior.* New York: Appleton-Century-Crofts.

Jordan, K. E., MacLean, E. L. and Brannon, E. M. (2008). Monkeys match and tally across senses. *Cognition, 108*, 617-625.

Judge, P. G., Evans, T. A. and Vyas, D. K. (2005). Ordinal representation of numeric quantities by brown capuchin monkeys (*Cebus apella*). *Journal of Experimental Psychology: Animal Behavior Processes, 31*, 79-94.

Luciano, M. and Nelson, C. A. (2002). Assessment of neuropsychological function through use of the Cambridge Neuropsychological Testing

Automated Battery: Performance in 4- to 12-year-old children. *Developmental Neuropsychology, 22*, 595-624.

McGonigle, B. J., Chalmers, M. and Dickinson, A. (2003). Concurrent disjoint and reciprocal classification by *Cebus apella* in seriation tasks: Evidence for hierarchical organization. *Animal Cognition, 6*, 185-197.

Maestripieri, D. E. (2003). *Primate psychology*. Cambridge, MA: Harvard University Press.

Mandrell, D. J. and Sackett, G. P. (2008). Comparability of developmental cognitive assessments between standard and computer testing methods. *Developmental Psychobiology, 51,* 1-13.

McClearn, G. E. and Harlow, H. F. (1954). The effect of spatial contiguity on discrimination learning by rhesus monkeys. *Journal of Comparative and Physiological Psychology, 45*, 391-394.

Menzel, C. R. (1999). Unprompted recall and reporting of hidden objects by a chimpanzee (*Pan troglodytes*) after extended delays. *Journal of Comparative Psychology, 113*, 426-434.

Meyer, D. M., Polidora, V. J. and McConnell, D. G. (1961). Effects of spatial S-R contiguity and response delay upon discriminative performances by monkeys. *Journal of Comparative and Physiological Psychology, 54*, 175-177.

Meyer, D. R., Treichler, F. R. and Meyer, P. M. (1965). Discrete-trial training techniques and stimulus variables. In A. M. Schrier and H. F. Harlow and F. Stollnitz (Eds.), *Behavior of nonhuman primates: Modern research trends (Vol 1)* (pp. 1-49). New York: Academic Press.

Morris, R. D. and Hopkins, W. D. (1993). Perception of human chimeric faces by chimpanzees: Evidence for a right hemisphere advantage. *Brain and Cognition, 21*, 111-122.

Mulcahy, N. J. and Call, J. (2006, 19 May). Apes save tools for future use. *Science, 312*, 1038-1040.

Murphy, J. V. and Miller, R. E. (1955). The effect of spatial contiguity of cue and reward in the object-quality learning of rhesus monkeys. *Journal of Comparative and Physiological Psychology, 48*, 221-224.

Murphy, J. V. and Miller, R. E. (1958). Effect of the spatial relationship between cue, reward, and response in simple discrimination learning. *Journal of Experimental Psychology, 56*, 26-31.

Nieder, A. and Merten, K. (2007). A labeled-line code for small and large numerosities in the monkey prefrontal cortex. *The Journal of Neuroscience, 27*, 5986-5993.

Olsen, J. (2008). Evaluating young children's cognitive capacities through computer versus hand drawings. *Scandinavian Journal of Psychology, 33*, 193-211.

Olthof, A., Iden, C. M. and Roberts, W. A. (1997). Judgments of ordinality and summation of number symbols by squirrel monkeys (*Saimiri sciureus*). *Journal of Experimental Psychology: Animal Behavior Processes, 23*, 325-339.

Olthof, A. and Roberts, W. A. (2000). Summation of symbols by pigeons (*Columba livia*): The importance of number and mass of reward items. *Journal of Comparative Psychology, 114*, 158-166.

Olthof, A. and Santi, A. (2007). Pigeons (*Columba livia*) associate time intervals with symbols in a touch screen task: Evidence for ordinality but not summation. *Journal of Comparative Psychology, 121*, 82-94.

Osvath, M. and Osvath, H. (2008). Chimpanzee (*Pan troglodytes*) and orangutan (*Pongo abelii*) forethought: Self-control and pre-experience in the face of future tool use. *Animal Cognition, 11*, 661-674.

Patterson, T. L. and Tzeng, O. J. L. (1979). Long-term memory for abstract concepts in the lowland gorilla (*Gorilla g. Gorilla*). *Bulletin of the Psychonomic Society, 13*, 279-282.

Reichmuth Kastak, C. and Schusterman, R. J. (2002). Long-term memory for concepts in a California sea lion (*Zalophus californianus*). *Animal Cognition, 5*, 225-232.

Richardson, W. K., Washburn, D. A., Hopkins, W. D., Savage-Rumbaugh, E. S. and Rumbaugh, D. M. (1990). The NASA/LRC Computerized Test System. *Behavior Research Methods, Instruments, and Computers, 22*, 127-131.

Rumbaugh, D. M. (1971). Evidence of qualitative differences in learning processes among primates. *Journal of Comparative and Physiological Psychology, 76*, 250-255.

Rumbaugh, D. M., King, J. E., Washburn, D. A., Beran, M. J. and Gould, K. L. (2007). A salience theory of learning and behavior - with perspectives on neurobiology and cognition. *International Journal of Primatology, 28*, 973-996.

Rumbaugh, D. M. and Pate, J. L. (1984a). The evolution of cognition in primates: A comparative perspective. In H. L. Roitblat and T. G. Bever and H. S. Terrace (Eds.), *Animal cognition* (pp. 569-587). Hillsdale, NJ: Erlbaum.

Rumbaugh, D. M. and Pate, J. L. (1984b). Primates' learning by levels. In G. Greenberg and E. Tobach (Eds.), *Behavioral evolution and integrative levels* (pp. 221-240). Hillsdale, NJ: Erlbaum.

Rumbaugh, D. M., Richardson, W. K., Washburn, D. A., Savage-Rumbaugh, E. S. and Hopkins, W. D. (1989). Rhesus monkeys (*Macaca mulatta*), video tasks, and implications for stimulus-response spatial contiguity. *Journal of Comparative Psychology, 103*, 32-38.

Santos, E. S., Hauser, M. D. and Spelke, E. S. (2001). Recognition and categorization of biologically significant objects by rhesus monkeys (*Macaca mulatta*): The domain of food. *Cognition, 82*, 127-155.

Schwartz, B. L., Hoffman, M. and Evans, S. (2005). Episodic-like memory in a gorilla: A review and new findings. *Learning and Motivation, 36*, 226-244.

Smith, B. R., Piel, A. K. and Candland, D. K. (2003). Numerity of a socially housed hamadryas baboon (*Papio hamadryas*) and a socially housed squirrel monkey (*Saimiri sciureus*). *Journal of Comparative Psychology, 117*, 217-225.

Smith, J. D., Beran, M. J., Redford, J. S. and Washburn, D. A. (2006). Dissociating uncertainty states and reinforcement signals in the comparative study of metacognition. *Journal of Experimental Psychology: General, 135*, 282-297.

Smith, J. D., Minda, J. P. and Washburn, D. A. (2004). Category learning in rhesus monkeys: A study of the Shepard, Hovland, and Jenkins (1961) task. *Journal of Experimental Psychology: General, 133*, 398-414.

Smith, J. D., Shields, W. E. and Washburn, D. A. (2003). A comparative approach to metacognition and uncertainty monitoring. *Behavioral and Brain Sciences, 26*, 317-339.

Spence, K. W. (1937). Analysis of the formation of visual discrimination habits in chimpanzees. *Journal of Comparative Psychology, 23*, 77-100.

Stahl, C. (2006). Software for generating psychological experiments. *Experimental Psychology, 53*, 218-232.

Suda-King, C. (2007). Do orangutans (*Pongo pygmaeus*) know when they do not remember? *Animal Cognition, 11*, 21-42.

Tanaka, M. (2001). Discrimination and categorization of photographs of natural objects by a chimpanzee (*Pan troglodytes*). *Animal Cognition, 4*, 201-211.

Thorndike, E. L. (1898). Animal intelligence: An experimental study of the associative processes in animals. *Psychological Review Monograph Supplement 2* (4, Whole No. 8).

Tomasello, M. and Call, J. (1997). *Primate cognition*. Oxford: Oxford University Press.

Tomonaga, M. (1998). Perception of shape from shading in chimpanzees (*Pan troglodytes)* and humans *(Homo sapiens*). *Animal Cognition, 1*, 25-35.

Tomonaga, M. (2007). Relative numerosity discrimination by chimpanzees (*Pan troglodytes*): Evidence for approximate numerical representations. *Animal Cognition, 11*, 43-57.

Vonk, J. and MacDonald, S. E. (2004). Levels of abstraction in orangutan (*Pongo abelii*) categorization. *Journal of Comparative Psychology, 118*, 3-13.

Washburn, D. A. (1990). PC-compatible computer-generated stimuli for video-task testing. *Behavior Research Methods, Instruments, and Computers, 22*, 132-135.

Washburn, D. A., Hopkins, W. D. and Rumbaugh, D. M. (1989). Video-task assessment of learning and memory in macaques (*Macaca mulatta*): Effects of stimulus movement on performance. *Journal of Experimental Psychology: Animal Behavior Processes, 15*, 393-400.

Washburn, D. A. and Rumbaugh, D. M. (1991a). Ordinal judgments of numerical symbols by macaques (*Macaca mulatta*). *Psychological Science, 2*, 190-193.

Washburn, D. A. and Rumbaugh, D. M. (1991b). Rhesus monkey (*Macaca mulatta*) complex learning skills reassessed. *International Journal of Primatology, 12*, 377-387.

Washburn, D. A. and Rumbaugh, D. M. (1992). Testing primates with joystick-based automated apparatus: Lessons from the Language Research Center's Computerized Test System. *Behavior Research Methods, Instruments, and Computers, 24*, 157-164.

Wasserman, E. A. and Zentall, T. R. E. (2006). *Comparative cognition: Experimental explorations of animal intelligence*. New York: Oxford University Press.

Zivotofsky, A. Z., Goldberg, M. E. and Powell, K. D. (2005). Rhesus monkeys behave as if they perceive the Duncker illusion. *Journal of Cognitive Neuroscience, 17*, 1011-1017.

Index

B

C

D

E

F

G

H

I

J

K

L

M

N

O

P

Q

R

S

T

U

V

W

Y

Z